Subharmonic Functions, Generalizations, Holomorphic Functions, Meromorphic Functions, and Properties

Authored by

Juhani Riihentaus

Department of Mathematical Sciences University of Oulu
P.O. Box 3000 FI-90014 Oulun yliopisto, Finland
&
Department of Physics and Mathematics
University of Eastern Finland P.O. Box 111
FI-80101 Joensuu
Finland

Subharmonic Functions, Generalizations, Holomorphic Functions, Meromorphic Functions, and Properties

Author: Juhani Riihentaus

ISBN (Online): 978-981-14-9870-1

ISBN (Print): 978-981-14-9868-8

ISBN (Paperback): 978-981-14-9869-5

Bentham Science Publishers Pte. Ltd.
80 Robinson Road #02-00
Singapore 068898
Singapore
Email: subscriptions@benthamscience.net

BENTHAM
SCIENCE

CONTENTS

PREFACE

Our presentation is divided into two parts. In the first part we consider subharmonic functions and their generalizations, so-called quasinearly subharmonic functions. In the second part we consider certain extension results for subharmonic functions, for holomorphic functions and for meromorphic functions.

Harmonic functions play a crucial role in mathematics. The same is true for a generalized class, for subharmonic functions. In this important area many authors, to mention just a few, Szpilrajn, Radó, Brelot, Lelong, Avanissian, Hervé, and Lieb and Loss, have found it useful to consider more general function classes, namely quasisubharmonic functions, nearly subharmonic functions, and almost subharmonic functions.

We are considering a rather general function class, namely quasinearly subharmonic functions. This class includes quasisubharmonic functions, nearly subharmonic functions and even almost subharmonic functions, at least more or less. Our class has its roots at least in the late fifties. The class of quasinearly subharmonic functions includes, in addition to nearly subharmonic functions, also functions satisfying certain growth conditions, especially certain eigenfunctions, polyharmonic functions, and subsolutions of certain general elliptic equations. Since harmonic functions are included in our class, nonnegative solutions of some elliptic equations are included. In particular, the partial differential equations associated with quasiregular mappings belong to this family of elliptic equations.

Though the class of quasinearly subharmonic functions is indeed large, the use of it seems, nevertheless, to be justified. In some instances, the use of quasinearly subharmonic functions makes it possible to simplify and clarify certain proofs of subharmonic functions, and sometimes even improve the existing results. As examples, the subharmonicity results of separately subharmonic functions are presented in section 5, and the weighted boundary behavior results of subharmonic functions in section 7.

In the second part we consider removability results for subharmonic functions, for separately subharmonic functions, for harmonic functions, for separately harmonic functions, and for holomorphic and for meromorphic functions. Our results are related, at least slightly, to the well-known already existing results of holomorphic and meromorphic functions.

CONSENT FOR PUBLICATION

Not applicable.

CONFLICT OF INTEREST

The author declares no conflict of interest, financial or otherwise.

ACKNOWLEDGEMENTS

Declared none.

Juhani Riihentaus
Department of Mathematical Sciences
University of Oulu
P.O. Box 3000
FI-90014 Oulun yliopisto, Finland and
Department of Physics and Mathematics
University of Eastern Finland
P.O. Box 111
FI-80101 Joensuu
Finland
E-mail: juhani.riihentaus@gmail.com

1. Quasinearly Subharmonic Functions

Abstract. We give the definition of quasinearly subharmonic functions, point out that this function class includes, among others, subharmonic functions, quasisubharmonic functions, nearly subharmonic functions (in a slightly generalized sense) and almost subharmonic functions (essentially). Moreover, we give basic properties of quasinearly subharmonic functions, and, among others, we characterize quasinearly subharmonicity with the aid of quasihyperbolic metric.

Keywords. Subharmonic, nearly subharmonic, quasinearly subharmonic, properties of quasinearly subharmonic functions, quasihyperbolic metric

1.1. Subharmonic functions and nearly subharmonic functions.

Let D be an open subset of $\mathbb{R}^N$, $N \geq 2$. We recall that an upper semicontinuous function $u : D \to [-\infty, +\infty)$ is *subharmonic* if for all $x \in D$ and $r > 0$ such that $\overline{B^N(x,r)} \subset D$,

$$u(x) \leq \frac{1}{\nu_N r^N} \int\limits_{B^N(x,r)} u(y)\, dm_N(y).$$

The function $u \equiv -\infty$ is considered subharmonic.

We say that a function $u : D \to [-\infty, +\infty)$ is *nearly subharmonic*, if u is Lebesgue measurable, $u^+ \in \mathcal{L}^1_{\text{loc}}(D)$, and for all $\overline{B^N(x,r)} \subset D$,

$$u(x) \leq \frac{1}{\nu_N r^N} \int\limits_{B^N(x,r)} u(y)\, dm_N(y). \tag{1.1}$$

Observe that in the standard definition of nearly subharmonic functions one uses the slightly stronger assumption that $u \in \mathcal{L}^1_{\text{loc}}(D)$, see e.g. [49], p. 14. However, our above, slightly more general definition seems to be more practical, see below Proposition 1.3 (iii) and Proposition 1.4 (vi) and (vii). Proceeding as in [49], proof of Theorem 1, pp. 14-15 (and referring also to these cited propositions), one gets the following result:

Lemma 1.1. *Let D be a domain in $\mathbb{R}^N$, $N \geq 2$. Let $u : D \to [-\infty, +\infty)$ be Lebesgue measurable. Then u is nearly subharmonic in D if and only if there exists a function u^*, subharmonic in D such that $u^* \geq u$ and $u^* = u$ Lebesgue almost everywhere in D. Here u^* is the lowest upper semicontinuous majorant of u:*

$$u^*(x) = \limsup_{x' \to x} u(x').$$

u^* is called the regularized subharmonic function to u.

Observe also that *almost subharmonic functions,* in the sense of Szpilrajn [147] (see also [88], 3.30, p. 20, and [68], p. 238 (Lieb and Loss even call this class briefly subharmonic functions)), are, more or less, included in our definition of nearly subharmonic functions, in the following sense. Let $u : D \to [-\infty, +\infty)$ be almost subharmonic, that is, $u \in \mathcal{L}^1_{\text{loc}}(D)$ and for Lebesgue almost every $x \in D$ and for every $r > 0$

Juhani Riihentaus

such that $\overline{B^N(x,r)} \subset D$ the mean value inequality (1.1) is satisfied. Let

$$D_1 := \{x \in D : u(x) \leq \frac{1}{\nu_N r^N} \int_{B^N(x,r)} u(y)\, dm_N(y) \text{ for all } \overline{B^N(x,r)} \subset D\}.$$

Define $\tilde{u} : D \to [-\infty, +\infty)$,

$$\tilde{u}(x) := \begin{cases} u(x), & \text{when } x \in D_1, \\ -\infty, & \text{when } x \in D \setminus D_1. \end{cases}$$

Since by assumption $m_N(D \setminus D_1) = 0$, it is easy to see that $\tilde{u}$ is nearly subharmonic in D.

In this connection see also [19].

1.2. **Quasinearly subharmonic functions.** Let D be a domain in $\mathbb{R}^N$, $N \geq 2$. It is an important fact that if $u : D \to [0, +\infty)$ is subharmonic and $p > 0$, then there exists a constant $C = C(N, p) > 0$ such that

$$u(x)^p \leq \frac{C}{\nu_N r^N} \int_{B^N(x,r)} u(y)^p\, dm_N(y) \tag{1.2}$$

for all $x \in D$ and $r > 0$ such that $\overline{B^N(x,r)} \subset D$. In the case $p = 1$, (1.2) is just the familiar mean value inequality with $C = 1$ for (nonnegative) subharmonic functions, (1.1) above. The case $p > 1$ follows from the the case $p = 1$ with the aid of Jensen's inequality. The case $0 < p < 1$ has been given in Fefferman and Stein [36], Lemma 2, p. 172, and by Kuran in [64], Theorem 1, p. 529, where, however, only absolute values of harmonic functions were considered. The proofs in [36, 64] are somewhat long. See also [38], Lemma 3.7, p. 116, and [2], (1.5), p. 210. In [94], Lemma, p. 69, it was pointed out that the proof of [36] includes the case of nonnegative subharmonic functions, too. See also [145], p. 271, [146], p. 114, [43], Lemma 1, p. 113, [140], Lemma 3, p. 305, [141], p. 794, [142], [143], Lemma 1, p. 363, [144], Lemma 2.1, p. 7, [26], Theorem A, p. 413, and [1], p. 132. Observe that a possibility for an essentially different proof was pointed out already in [149], pp. 188-190. Later other different proofs were given in [80], p. 18 and Theorem 1, p. 19 (see also [81], Theorem A, p. 15), [97], pp. 233-234, and [99], p. 188. The results in [80, 97, 99] hold in fact for more general function classes than just for nonnegative subharmonic functions. Compare also [22], [27], p. 430, and [28], p. 485.

As already pointed out, the inequality (1.2) is indeed useful. It has many applications. Among others, it has been applied to the weighted boundary behavior of subharmonic functions, to the nonintegrability of subharmonic and superharmonic functions (see section 2 below and the references therein) and especially to the subharmonicity of separately subharmonic functions (see section 5 below and the references therein).

It is therefore relevant to try find generalizations or related results to the inequality (1.2). We will do this in the following way.

We say that a Lebesgue measurable function $u : D \to [-\infty, +\infty)$ is *K-quasinearly subharmonic*, if $u^+ \in \mathcal{L}^1_{\text{loc}}(D)$ and if there is a constant $K = K(N, u, D) \geq 1$ such that for all $x \in D$ and $r > 0$ such that $\overline{B^N(x,r)} \subset D$, one has

$$u_M(x) \leq \frac{K}{\nu_N r^N} \int\limits_{B^N(x,r)} u_M(y)\, dm_N(y)$$

for all $M \geq 0$, where $u_M := \max\{u, -M\} + M$. A function $u : D \to [-\infty, +\infty)$ is *quasinearly subharmonic*, if u is K-quasinearly subharmonic in D for some $K \geq 1$.

In addition to the above defined class of quasinearly subharmonic functions, we will, but only in passing, consider a proper subclass, too. A Lebesgue measurable function $u : D \to [-\infty, +\infty)$ is *K-quasinearly subharmonic n.s. (in the narrow sense)*, if $u^+ \in \mathcal{L}^1_{\text{loc}}(D)$ and if there is a constant $K = K(N, u, D) \geq 1$ such that for all $\overline{B^N(x,r)} \subset D$, one has

$$u(x) \leq \frac{K}{\nu_N r^N} \int\limits_{B^N(x,r)} u(y)\, dm_N(y).$$

A function $u : D \to [-\infty, +\infty)$ is *quasinearly subharmonic n.s.*, if u is K-quasinearly subharmonic n.s. in D for some $K \geq 1$. Observe that if $u : D \to [0, +\infty)$ is subharmonic and $p > 0$, then u^p is quasinearly subharmonic n.s. and thus also quasinearly subharmonic, see below Proposition 1.3 (i) and (iii). See also [31].

More generally, the class of quasinearly subharmonic functions includes, among others, subharmonic functions, and, more generally, quasisubharmonic and nearly subharmonic functions (for the definitions of these, see above, [49, 109], also functions satisfying certain natural growth conditions, especially certain eigenfunctions, polyharmonic functions, subsolutions of certain general elliptic equations. Also, the class of Harnack functions is included, thus, among others, nonnegative harmonic functions as well as nonnegative solutions of some elliptic equations. In particular, the partial differential equations associated with quasiregular mappings belong to this family of elliptic equations, see Vuorinen [151] and the above references.

Observe that already Domar [27] has pointed out the relevance of the class of (nonnegative) quasinearly subharmonic functions. For, at least partly, an even more general function class, see [28].

To motivate the above defined function classes, the class of quasinearly subharmonic functions and the class of quasinearly subharmonic n.s. functions, even more, we point out just the following. First, in order to see whether, for a property of subharmonic functions, subharmonicity is really essential or not, just check whether the property in question does hold for quasinearly subharmonic functions or not. Second, and on the other hand, combining results of quasinearly subharmonic functions with results of subharmonic functions, one can, at least in some cases, improve results of subharmonic functions. One such an example is our improvement of Armitage's and Gardiner's separate subharmonicity result [3], Theorem 1, p. 256. As a matter of fact, our improvement, Corollary 5.7 and Example 11 below, follows from Theorem 5.5 (a result for separately quasinearly subharmonic functions (which does not hold in the subharmonic case)) and from [94], Theorem 1, p. 69 ([113], Theorem 3B, p. e2620), a result of separately subharmonic functions. For another such an example (at least partly), see a short and direct proof for [109], Corollary 3.3, pp. 62-63 ([113], Corollary 3.2.5, p. e2621), below in Corollary 5.14.

Below we list four simple examples of quasinearly subharmonic functions.

Example 1. Let D be a domain in $\mathbb{R}^N$, $N \geq 2$. Any Lebesgue measurable function $u : D \to [m, M]$, where $0 < m \leq M < +\infty$, is quasinearly subharmonic, and, because of Proposition 1.3 (i) (see below), also quasinearly subharmonic n.s. If u is moreover continuous, it is even a Harnack function, see [151], pp. 259, 263.

Example 2. The function $u : \mathbb{R}^2 \to \mathbb{R}$,

$$u(x,y) := \begin{cases} -1, & \text{when } y < 0, \\ 1, & \text{when } y \geq 0, \end{cases}$$

is 2-quasinearly subharmonic, but not quasinearly subharmonic n.s.

Example 3. Let $D = (0,2) \times (0,1) \subset \mathbb{R}^2$, let $c < 0$ be arbitrary. Let $E \subset D$ be a Borel set of zero Lebesgue measure. Let $u : D \to [-\infty, +\infty)$,

$$u(x,y) := \begin{cases} c, & \text{when } (x,y) \in E, \\ 1, & \text{when } (x,y) \in D \setminus E \text{ and } 0 < x < 1, \\ 2, & \text{when } (x,y) \in D \setminus E \text{ and } 1 \leq x < 2. \end{cases}$$

The function u attains both negative and positive values, is 2-quasinearly subharmonic n.s, but not nearly subharmonic.

Example 4. Let D be a domain in $\mathbb{R}^N$, $N \geq 2$, and let $u : D \to [-\infty, +\infty)$ be any quasinearly subharmonic function. Let $E \subset D$ be a Borel set of zero Lebesgue measure. Let $v : D \to [-\infty, +\infty)$,

$$v(x) := \begin{cases} -\infty, & \text{when } x \in E, \\ u(x), & \text{when } x \in D \setminus E. \end{cases}$$

The function v is clearly quasinearly subharmonic.

Quasinearly subharmonic functions (perhaps with a different terminology), or, essentially, perhaps just functions satisfying a certain generalized mean value inequality, have previously been considered, or used, in addition to the references listed already above or below, at least in [74, 101, 102, 103, 106, 107, 25, 60, 111, 112, 29, 30, 63, 72].

1.3. **Basic properties of quasinearly subharmonic functions.** We begin by defining so called *permissible* functions. A function $\psi : [0,+\infty) \to [0,+\infty)$ is *permissible*, if there exist an increasing (strictly or not), convex function $\psi_1 : [0,+\infty) \to [0,+\infty)$ and a strictly increasing surjection $\psi_2 : [0,+\infty) \to [0,+\infty)$ such that $\psi = \psi_2 \circ \psi_1$ and such that the following conditions are satisfied:

(a) ψ_1 satisfies the Δ_2-condition,
(b) ψ_2^{-1} satisfies the Δ_2-condition,
(c) the function $t \mapsto \frac{\psi_2(t)}{t}$ is *quasi-decreasing*, i.e. there is a constant $C = C(\psi_2) > 0$ such that

$$\frac{\psi_2(s)}{s} \geq C\frac{\psi_2(t)}{t}$$

for all $0 \leq s \leq t$.

A function $\psi : [0,+\infty) \to [0,+\infty)$ is *strictly permissible*, if it is an increasing surjection satisfying the conditions (b) and (c).

Recall that a function $\varphi : [0,+\infty) \to [0,+\infty)$ satisfies a Δ_2-*condition*, if there is a constant $C = C(\varphi) \geq 1$ such that $\varphi(2t) \leq C\varphi(t)$ for all $t \in [0,+\infty)$.

Permissible functions are necessarily continuous.

Examples of permissible functions are: $\psi_1(t) = t^p$, $p > 0$, and $\psi_2(t) = ct^{p\alpha}[\log(\delta + t^{p\gamma})]^\beta$, $c > 0$, $0 < \alpha < 1$, $\delta \geq 1$, $\beta, \gamma \in \mathbb{R}$ such that $0 < \alpha + \beta\gamma < 1$, and $p \geq 1$. And also functions of the form $\psi_3 = \phi \circ \varphi$, where $\phi : [0,+\infty) \to [0,+\infty)$ is a concave surjection whose inverse ϕ^{-1} satisfies the Δ_2-condition and $\varphi : [0,+\infty) \to [0,+\infty)$ is an increasing, convex function satisfying the Δ_2-condition.

Though permissible functions are more general than nonnegative power functions, the following lemma shows that in most cases it indeed suffices to work with power functions:

Lemma 1.2. ([85], Lemma 1 and Remark 1, p. 93) *Let $\psi : [0,+\infty) \to [0,+\infty)$ be a permissible function. Then*

(i) *there are a number $p > 0$ and a convex function $M : [0,+\infty) \to [0,+\infty)$ satisfying the Δ_2-condition such that $\psi(t) \asymp M(t^p)$, that is, there exist constants $C_1 > 0$ and $C_2 > 0$ such that*

$$C_1 \leq \frac{\psi(t)}{M(t^p)} \leq C_2$$

for all $t > 0$,

(ii) *there are a number $p > 0$ and a convex function $\vartheta : [0,+\infty) \to [0,+\infty)$ satisfying the Δ_2-condition such that $\psi(t) \asymp \vartheta(t)^p$.*

Next we list certain basic properties of quasinearly subharmonic functions, see [109], Proposition 2.1 and Proposition 2.2, pp. 54-55, and [113], Proposition 1.5.1 and Proposition 1.5.2, p. e2615.

Proposition 1.3. *Let D be a domain in $\mathbb{R}^N$, $N \geq 2$.*

(i) *If* $u : D \to [0,+\infty)$ *is Lebesgue measurable and* $u^+ \in \mathcal{L}^1_{\mathrm{loc}}(D)$ *, then* u *is* K*-quasinearly subharmonic if and only if* u *is* K*-quasinearly subharmonic n.s., that is, if for all* $\overline{B^N(x,r)} \subset D$,

$$u(x) \leq \frac{K}{\nu_N r^N} \int\limits_{B^N(x,r)} u(y)\, dm_N(y).$$

(ii) *If* $u : D \to [-\infty,+\infty)$ *is* K*-quasinearly subharmonic n.s., then* u *is* K*-quasinearly subharmonic in* D*, but not necessarily conversely.*

(iii) *A function* $u : D \to [-\infty,+\infty)$ *is* 1*-quasinearly subharmonic if and only if it is nearly subharmonic, that is, it is* 1*-quasinearly subharmonic n.s.*

(iv) *If* $u : D \to [0,+\infty)$ *is quasinearly subharmonic and* $\psi : [0,+\infty) \to [0,+\infty)$ *is permissible, then* $\psi \circ u$ *is quasinearly subharmonic in* D*. Especially, if* $h : D \to \mathbb{R}$ *is harmonic and* $p > 0$*, then* $|\, h\,|^p$ *is quasinearly subharmonic.*

(v) *Harnack functions are quasinearly subharmonic.*

Proof. We leave the cases (i), (ii) and (v) to the reader. For the proof of (iv), see [105], Lemma 2.1, p. 32. To prove the case (iii) suppose that u is nearly subharmonic in D. Then clearly u_M is nearly subharmonic for all $M \geq 0$, and thus for all $x \in D$ and $r > 0$ such that $\overline{B^N(x,r)} \subset D$, one has

$$u_M(x) \leq \frac{1}{\nu_N r^N} \int\limits_{B^N(x,r)} u_M(y)\, dm_N(y).$$

Hence u is 1-quasinearly subharmonic.

On the other hand, if u is 1-quasinearly subharmonic in D, then one sees at once, with the aid of the Lebesgue Monotone Convergence Theorem, that u is nearly subharmonic in D. □

Proposition 1.4. *Let* D *be a domain in* $\mathbb{R}^N$, $N \geq 2$.

(i) *If* $u : D \to [-\infty,+\infty)$ *is* K_1*-quasinearly subharmonic and* $K_2 \geq K_1$*, then* u *is* K_2*-quasinearly subharmonic in* D*.*

(ii) *If* $u_1 : D \to [-\infty,+\infty)$ *and* $u_2 : D \to [-\infty,+\infty)$ *are* K*-quasinearly subharmonic n.s., then* $\lambda_1 u_1 + \lambda_2 u_2$ *is* K*-quasinearly subharmonic n.s. in* D *for all* $\lambda_1, \lambda_2 \geq 0$.

(iii) *If* $u : D \to [-\infty,+\infty)$ *is quasinearly subharmonic, then* u *is locally bounded above in* D*.*

(iv) *If* $u_j : D \to [-\infty,+\infty)$, $j = 1,2,\ldots$, *are* K*-quasinearly subharmonic (resp.* K*-quasinearly subharmonic n.s.), and* $u_j \searrow u$ *as* $j \to +\infty$*, then* u *is* K*-quasinearly subharmonic (resp.* K*-quasinearly subharmonic n.s.) in* D*.*

(v) *If* $u : D \to [-\infty,+\infty)$ *is* K_1*-quasinearly subharmonic and* $v : D \to [-\infty,+\infty)$ *is* K_2*-quasinearly subharmonic, then* $\max\{u,v\}$ *is* $\max\{K_1,K_2\}$*-quasinearly subharmonic in* D*. Especially,* $u^+ := \max\{u,0\}$ *is* K_1*-quasinearly subharmonic in* D*.*

(vi) *Let $\mathcal{F}$ be a family of K-quasinearly subharmonic (resp. K-quasinearly subharmonic n.s.) functions in D and let $w := \sup_{u\in\mathcal{F}} u$. If w is Lebesgue measurable and $w^+ \in \mathcal{L}^1_{\mathrm{loc}}(D)$, then w is K-quasinearly subharmonic (resp. K-quasinearly subharmonic n.s.) in D.*

(vii) *If $u : D \to [-\infty,+\infty)$ is quasinearly subharmonic n.s., then either $u \equiv -\infty$ or u is finite almost everywhere in D, and $u \in \mathcal{L}^1_{\mathrm{loc}}(D)$.*

Remark 1.5. Related to (ii) above, it is easy to see that, if $u : D \to [-\infty,+\infty)$ is K-quasinearly subharmonic, then $\lambda u + C$ is K-quasinearly subharmonic in D for all $\lambda \geq 0$ and $C \geq 0$.

However, the sum of quasinearly subharmonic functions is not necessarily quasinearly subharmonic. This is seen from [32], Example 5, p. 251. As a matter of fact, let $u : \mathbb{R}^2 \to \mathbb{R}$,

$$u(x,y) := \begin{cases} 3, & \text{when } x = 0, \\ 1, & \text{when } x \neq 0, \end{cases}$$

which is 3-quasinearly subharmonic. The constant function $v : \mathbb{R}^2 \to \mathbb{R}$, $v(x,y) \equiv -2$ is harmonic. One sees easily that $u+v : \mathbb{R}^2 \to \mathbb{R}$,

$$(u+v)(x,y) = \begin{cases} 1, & \text{when } x = 0, \\ -1, & \text{when } x \neq 0, \end{cases}$$

is not quasinearly subharmonic.

Proof. We leave the rather easy cases (i)–(vi) to the reader and prove only (vii). Our proof is nearly verbatim the same as [48], proof of Theorem 4.10, p. 66, except perhaps in a couple of the last lines of the proof.

By (vi) u^+ is quasinearly subharmonic n.s., and by (iii) locally bounded above in D. Therefore, for each $\overline{B^n(x,r)} \subset D$, the integral

$$\int\limits_{B^N(x,r)} u(y)\,dm_N(y)$$

is defined either as $-\infty$ or as a finite real number. Suppose that $u(x_0) > -\infty$ for some $x_0 \in D$. Then

$$-\infty < u(x_0) \leq \frac{K}{\nu_N r^N} \int\limits_{B^N(x_0,r)} u(y)\,dm_N(y) < +\infty.$$

Therefore there exists $r_0 > 0$ such that $u(x) \in \mathbb{R}$ for almost every $x \in B^N(x_0,r_0)$.

Write

$$H := \{x \in D : u(y) \text{ finite for a.e. } y \in B^N(x,r) \text{ for some } r > 0 \text{ s.t. } \overline{B^N(x,r)} \subset D\}.$$

From above it follows that $H \neq \emptyset$. It is easy to see that H is open. To show that H is closed in D, take a sequence $x_j \in H$, $j = 1,2,\ldots$, $x_j \to x^*$ as $j \to +\infty$. Take $r^* > 0$ such that $\overline{B^N(x^*,r^*)} \subset D$. Choose $x_{j_0} \in H \cap B^N(x^*,\frac{r^*}{2})$. Then there is $r_{j_0} > 0$ such that $u(x) \in \mathbb{R}$ for almost every $x \in B^N(x_{j_0},r_{j_0})$. Hence also $u(x) \in \mathbb{R}$ for almost every

$x \in B^N(x_{j_0}, r_{j_0}) \cap B^N(x^*, \frac{r^*}{2})$. Let A be the set of such points. Clearly A is of positive measure. Choose $\hat{x} \in A$. Then

$$-\infty < u(\hat{x}) \leq \frac{K}{\nu_N \left(\frac{r^*}{2}\right)^N} \int_{B^N(\hat{x}, \frac{r^*}{2})} u(y)\, dm_N(y) < +\infty.$$

Hence $u(x) \in \mathbb{R}$ for almost every $x \in B^N(\hat{x}, \frac{r^*}{2})$. On the other hand, $x^* \in B^N(\hat{x}, \frac{r^*}{2})$. Thus x^* has a neighborhood $B^N(x^*, \delta) \subset B^N(\hat{x}, \frac{r^*}{2})$ such that $u(x) \in \mathbb{R}$ for almost every $x \in B^N(x^*, \delta)$. Therefore $x^* \in H$. Since D is connected, $H = D$.

To show that $u \in \mathcal{L}^1_{\mathrm{loc}}(D)$, take $\overline{B^N(x_0, r_0)} \subset D$ arbitrarily. We know that for each $x \in \overline{B^N(x_0, r_0)}$ there is $r_x > 0$ such that $u(y) \in \mathbb{R}$ for almost every $y \in B^N(x, r_x)$. Then

$$B^N(x, r_x), \quad x \in \overline{B^N(x_0, r_0)},$$

is an open cover of the compact set $\overline{B^N(x_0, r_0)}$. Hence we find a finite subcover

$$B^N(x_1, r_1), B^N(x_2, r_2), \ldots, B^N(x_k, r_k).$$

We may suppose that $u(x_j) \in \mathbb{R}$ for all $j = 1, 2, \ldots, k$. One achieves this just replacing x_j, if necessary, by a nearby point x_j^*, and possibly increasing r_j a little bit, $j = 1, 2, \ldots, k$ (which is indeed possible). Then

$$-\infty < u(x_j) \leq \frac{K}{\nu_N r_j^N} \int_{B^N(x_j, r_j)} u(y) dm_N(y), \quad j = 1, 2, \ldots, k.$$

By (iii) there is a constant $C > 0$ such that $u(y) \leq u^+(y) \leq C < +\infty$ for each $y \in \overline{B^N(x_1, r_1)} \cup \overline{B^N(x_2, r_2)} \cup \cdots \cup \overline{B^N(x_k, r_k)}$. Therefore $u(y) - C \leq 0$ for each $y \in \overline{B^N(x_1, r_1)} \cup \overline{B^N(x_2, r_2)} \cup \cdots \cup \overline{B^N(x_k, r_k)}$, and we get

$$\begin{aligned}
-\infty < \sum_{j=1}^{k} \nu_N r_j^N u(x_j) &\leq \sum_{j=1}^{k} K \cdot \int_{B^N(x_j, r_j)} u(y) dm_N(y) \leq \\
&\leq K \cdot \sum_{j=1}^{k} \int_{B^N(x_j, r_j)} [u(y) - C] dm_N(y) + K \cdot C \cdot \sum_{j=1}^{k} m_N(B^N(x_j, r_j)) \leq \\
&\leq K \cdot \int_{B^N(x_0, r_0)} [u(y) - C] dm_N(y) + K \cdot C \cdot \sum_{j=1}^{k} m_N(B^N(x_j, r_j)) \leq \\
&\leq K \cdot \int_{B^N(x_0, r_0)} u(y) dm_N(y) - K \cdot C m_N(B^N(x_0, r_0)) + K \cdot C \cdot \sum_{j=1}^{k} m_N(B^N(x_j, r_j)) < \\
&< +\infty.
\end{aligned}$$

Thus

$$-\infty < \int\limits_{B^N(x_0,r_0)} u(y)\, dm_N(y) < +\infty,$$

and the claim follows. □

Remark 1.6. It is easy to see that (vii) does not anymore hold for quasinearly subharmonic functions. As a counterexample serves the function $u : \mathbb{R}^2 \to [-\infty, +\infty)$,

$$u(x,y) := \begin{cases} -\infty, & \text{when } y \leq 0, \\ 1, & \text{when } y > 0, \end{cases}$$

which is 2-quasinearly subharmonic, but surely not quasinearly subharmonic n.s.

1.4. **Quasinearly subharmonic functions and quasihyperbolic metric.** We characterize nonnegative quasinearly subharmonic functions with the aid of the quasihyperbolic metric. For this purpose we recall the definitions of the quasihyperbolic distance and volume, see e.g. Vuorinen [152], p. 33.

Let D be a proper subdomain of $\mathbb{R}^N$, $N \geq 2$. The *quasihyperbolic distance* between x and y in D is defined by

$$k_D(x,y) = \inf_{\alpha \in \Gamma_{xy}} \int_\alpha \frac{1}{\operatorname{dist}(x, \partial D)} \mid dx \mid,$$

where

$$\Gamma_{xy} = \{\alpha : \alpha \text{ is a rectifiable curve joining } x \text{ and } y \text{ in } D\}.$$

It is clear that k_D is a metric on D. Write

$$B_q(x,r) := \{y \in D : k_D(x,y) < r\}.$$

Let $R > 0$, and $\rho = 1 - e^{-R}$, $\hat{\rho} = e^R - 1$. Then by [152], (3.9), p. 35,

$$B^N(x, \rho\delta(x)) \subset B_q(x,R) \subset B^N(x, \hat{\rho}\delta(x)). \tag{1.3}$$

The *quasihyperbolic volume* of a Lebesgue measurable set $A \subset D$ is defined by

$$m_q(A) = \int_A \frac{1}{\operatorname{dist}(x, \partial D)^N}\, dm_N(x),$$

where dm_N is the standard, non-normalized Lebesgue measure in $\mathbb{R}^N$.

Theorem 1.7. *Let D be a proper subdomain of $\mathbb{R}^N$, $N \geq 2$. Suppose that $u : D \to [-\infty, +\infty)$ and $u^+ \in \mathcal{L}^1_{\mathrm{loc}}(D)$. Then u is quasinearly subharmonic if and only if there is a constant $C > 0$ such that for all $M \geq 0$,*

$$u_M(x) \leq \frac{C}{m_q(B_q(x,R))} \int_{B_q(x,R)} u_M(y)\, dm_q(y)$$

for all $x \in D$ and R, $0 < R \leq \log\frac{4}{3}$. Here $u_M = \max\{u, -M\} + M$.

Proof. Consider first the case when u is quasinearly subharmonic. Let R, ρ and $\hat{\rho}$ be as above. Suppose moreover that $0 < R \leq \log\frac{4}{3}$, thus $\rho \leq \frac{1}{4}$ and $\hat{\rho} \leq \frac{1}{3}$. Let $M \geq 0$ be arbitrary. Since u is quasinearly subharmonic,

$$\begin{aligned}
u_M(x) &\leq \frac{C}{m_N(B^N(x,\rho\,\delta(x)))}\int_{B^N(x,\rho\,\delta(x))} u_M(y)\,dm_N(y) = \\
&\leq \frac{C}{m_N(B^N(x,\rho\,\delta(x)))}\int_{B^N(x,\rho\,\delta(x))} u_M(y)\,\mathrm{dist}(y,\partial D)^N \frac{dm_N(y)}{\mathrm{dist}(y,\partial D)^N} = \\
&\leq \frac{C}{m_N(B^N(x,\hat{\rho}\,\delta(x)))}\cdot\frac{m_N(B^N(x,\hat{\rho}\,\delta(x)))}{m_N(B^N(x,\rho\,\delta(x)))}\int_{B_q(x,R)} u_M(y)\,\mathrm{dist}(y,\partial D)^N\,dm_q(y) \leq \\
&\leq \frac{C}{[(1-\hat{\rho})\delta(x)]^N\, m_q(B_q(x,R))}\cdot\left(\frac{\hat{\rho}}{\rho}\right)^N\cdot[(1+\hat{\rho})\delta(x)]^N\int_{B_q(x,R)} u_M(y)\,dm_q(y) \leq \\
&\leq \left(\frac{1+\hat{\rho}}{1-\hat{\rho}}\right)^N\cdot\left(\frac{4}{3}\right)^N\cdot\frac{C}{m_q(B_q(x,R))}\int_{B_q(x,R)} u_M(y)\,dm_q(y) \leq \\
&\leq \frac{C}{m_q(B_q(x,R))}\int_{B_q(x,R)} u_M(y)\,dm_q(y).
\end{aligned}$$

Above we have used (1.3), the facts that $m_N(B^N(x,r)) = \nu_N r^N$, that $0 < R \leq \log\frac{4}{3}$, $\rho \leq \frac{1}{4}$ and $\hat{\rho} \leq \frac{1}{3}$, and that the inequalities

$$(1-\hat{\rho})\delta(x) \leq \mathrm{dist}(y,\partial D) \leq (1+\hat{\rho})\delta(x),$$

hold for all $y \in B^n(x,\hat{\rho}\delta(x))$, thus

$$m_N(B^N(x,\hat{\rho}\delta(x))) \geq [(1-\hat{\rho})\delta(x)]^N m_q(B_q(x,R)).$$

To prove the if case, suppose that u is quasinearly subharmonic *in the quasihyperbolic sense*, i.e. $u^+ \in \mathcal{L}^1_{\mathrm{loc}}(D)$ and that there is a constant $C > 0$ such that for all $M \geq 0$

$$u_M(x) \leq \frac{C}{m_q(B_q(x,R))}\int_{B_q(x,R)} u_M(y)\,dm_q(y)$$

for all $x \in D$ and R, $0 < R < \log\frac{4}{3}$. We must show that

$$u_M(x) \leq \frac{C}{m_N(B^N(x,\hat{R}\delta(x)))}\int_{B^N(x,\hat{R}\delta(x))} u_M(y)\,dm_N(y) \tag{1.4}$$

for all $\hat{R}$, $0 < \tilde{R} < 1$.

We suppose first that $0 < \hat{R} \leq \log\frac{4}{3}$. Using similar estimates as above, e.g. the fact that

$$m_q(B_q(x,\hat{R})) \geq \frac{1}{[(1+\hat{\rho})\delta(x)]^N}\, m_N(B^N(x,\rho\delta(x))),$$

we get

$$\begin{aligned}
u_M(x) &\leq \frac{C}{m_q(B_q(x,\hat{R}))}\int_{B_q(x,\hat{R})} u_M(y)\,dm_q(y) \leq \\
&\leq \frac{C}{m_N(B^N(x,\rho\,\delta(x)))}\cdot[(1+\hat{\rho})\delta(x)]^N \int_{B^N(x,\hat{\rho}\,\delta(x))} u_M(y)\,\frac{dm_N(y)}{\operatorname{dist}(y,\partial D)^N} \leq \\
&\leq \frac{C}{m_N(B^N(x,\rho\,\delta(x)))}\cdot\frac{[(1+\hat{\rho})\delta(x)]^N}{[(1-\hat{\rho})\delta(x)]^N}\int_{B^N(x,\hat{\rho}\delta(x))} u_M(y)\,dm_N(y) \leq \\
&\leq \frac{C}{m_N(B^N(x,\rho\,\delta(x)))}\cdot\left(\frac{1+\hat{\rho}}{1-\hat{\rho}}\right)^N \int_{B^N(x,\hat{\rho}\delta(x))} u_M(y)\,dm_N(y) \leq \\
&\leq \frac{C}{m_N(B^N(x,\hat{R}\delta(x)))}\cdot\frac{m_N(B^N(x,\hat{R}\delta(x)))}{m_N(B^N(x,\rho\delta(x)))}\cdot 2^N \int_{B^N(x,\hat{R}\delta(x))} u_M(y)\,dm_N(y) \leq \\
&\leq \frac{C}{m_N(B^N(x,\hat{R}\delta(x)))}\cdot\left(\frac{\hat{R}}{\rho}\right)^N\cdot 2^N \int_{B^N(x,\hat{R}\delta(x))} u_M(y)\,dm_N(y).
\end{aligned}$$

If $\hat{R}\leq\frac{1}{3}$, choose above $\hat{\rho}=\hat{R}$. Then $\frac{\hat{R}}{\rho}\leq\frac{4}{3}$, and the claim (1.4) follows. If $\frac{1}{3}<\hat{R}<1$, then choosing above $\rho=\frac{1}{4}$, say, one gets the claim. □

2. Modifications of the Mean Value Inequality for Quasinearly Subharmonic Functions

Abstract. We find necessary and sufficient conditions under which subsets of balls are big enough for the characterization of nonnegative, quasinearly subharmonic functions by mean value inequalities. A similar result is obtained also for generalized mean value inequalities where, instead of balls, we consider arbitrary bounded sets which have nonvoid interiors and instead of the volume of ball we use functions depending on the radius of this ball.

Keywords. Subharmonic, nearly subharmonic, quasinearly subharmonic, mean value

2.1. **Definitions and related results.** Let $\Omega \subseteq \mathbb{R}^N$ be an open set. We recall that a Lebesgue measurable function $u : \Omega \to [0, +\infty)$ is K-quasinearly subharmonic if and only if $u \in \mathcal{L}^1_{\mathrm{loc}}(\Omega)$ and

$$u(x) \leq \frac{K}{m_N(B^N(x,r))} \int\limits_{B^N(x,r)} u(y)\, dm_N(y) \tag{2.1}$$

for all closed balls $\overline{B^N(x,r)} \subset \Omega$.

We will relax, in a certain sense, the mean value inequality (2.1). As a matter of fact, we present certain necessary and sufficient conditions under which subsets of balls are big enough for the characterization of nonnegative quasinearly subharmonic functions by mean value inequalities.

We write $QNS(\Omega)$ for the set of all quasinearly subharmonic functions on an open set $\Omega \subseteq \mathbb{R}^N$, and $QNS_+(\Omega)$ for the set of all nonnegative quasinearly subharmonic functions on an open set $\Omega \subseteq \mathbb{R}^N$.

Observe that that if $u \in QNS_+(\Omega)$, then (2.1) holds also for every open ball $B^N(x,r) \subseteq \Omega$.

Let A be a subset of the open half-line $(0, +\infty)$ such that 0 is a limit point of A. Let $u : \Omega \to [-\infty, +\infty)$ be an upper semicontinous function on an open set $\Omega \subseteq \mathbb{R}^N$. The classical Blaschke–Privalov theorem, see, for example, [11], Chapter II, implies that u is subharmonic if

$$u(x) \leq \frac{1}{\nu_N r^N} \int\limits_{B^N(x,r)} u(y)\, dm_N(y) \tag{2.2}$$

holds whenever $r \in A$ and $\overline{B^N(x,r)} \subset \Omega$. Moreover, simple examples show that if nonnegative $u \in \mathcal{L}^1_{\mathrm{loc}}(\Omega)$, then the fulfilment of (2.1) for all $(x,r) \in \Omega \times A$ with $\overline{B^N(x,r)} \subset \Omega$ does not, generally, imply $u \in QNS(\Omega)$. A legitimate question to raise at this point is in finding sets $A \subseteq (0, +\infty)$ for which every nonnegative $u \in \mathcal{L}^1_{\mathrm{loc}}(\Omega)$ is quasinearly subharmonic if (2.1) holds for $(x,r) \in \Omega \times A$ whenever $\overline{B^N(x,r)} \subset \Omega$.

Juhani Riihentaus

Definition 2.1. Let Ω be an open set in $\mathbb{R}^N$. A set $A \subseteq (0, +\infty)$ is *favorable for Ω (favorable for the characterization of nonnegative, quasinearly subharmonic functions in Ω)* if for every nonnegative $u \in \mathcal{L}^1_{\text{loc}}(\Omega)$ the following conditions are equivalent:

(a) $u \in QNS(\Omega)$.

(b) There is $K = K(u, A, \Omega) \geq 1$ such that for all $x \in \Omega$ the inequality

$$u(x) \leq \frac{K}{\nu_N r^N} \int_{B^N(x,r)} u(y)\, dm_N(y) \tag{2.3}$$

holds whenever $r \in A$ and $\overline{B^N(x,r)} \subset \Omega$.

We can characterize the favorable subsets of $(0, +\infty)$ in the following way.

Theorem 2.2. ([31], Theorem 1.5, p. 351) *The following three statements are equivalent for every $A \subseteq (0, +\infty)$:*

(i) *A is favorable for all open sets $\Omega \subseteq \mathbb{R}^N$.*

(ii) *The characteristic function*

$$\chi_\Gamma(x) = \begin{cases} 1 & \text{if } x \in \Gamma, \\ 0 & \text{if } x \in \Omega \setminus \Gamma \end{cases}$$

is quasinearly subharmonic for all open sets $\Omega \subseteq \mathbb{R}^N$ and all Lebesgue measurable sets $\Gamma \subseteq \Omega$ if and only if there is a constant $K = K(\Gamma, \Omega, N)$ such that the inequality

$$m_N(B^N(x,r)) \leq K m_N(\Gamma \cap B^N(x,r)) \tag{2.4}$$

holds for $(x, r) \in \Omega \times A$ whenever $\overline{B^N(x,r)} \subset \Omega$.

(iii) *There exists $C = C(A) > 1$ such that*

$$\left[\frac{x}{C}, x\right] \cap A \neq \emptyset$$

for every $x \in (0, +\infty)$.

We shall prove the equivalence (i)$\Leftrightarrow$(iii) in Theorem 2.4 below. Observe also that the implication (i)$\Rightarrow$(ii) is trivial and that (ii)$\Rightarrow$(iii) follows directly from the proof of Theorem 2.3.

Quasidisks give an important example of sets Γ such that (2.4) holds in a bounded domain $\Omega \subseteq \mathbb{R}^2$ whenever $\overline{B^2(x,r)} \subset \Omega$. It is a particular case of the Gehring–Martio result which proves (2.4) for the so-called quasiextremal distance domains in $\mathbb{R}^N$, $N \geq 2$. See [40], Lemma 2.13.

The following result, closely connected to Theorem 2.2, follows from Theorem 2.22, see below.

Theorem 2.3. ([31], Theorem 1.6, pp. 351-352) *Let f be a positive function on $(0, +\infty)$. The following three statements are equivalent:*

(i) *For all open sets* $\Omega \subseteq \mathbb{R}^N$, *Lebesgue measurable functions* $u : \Omega \to [0, +\infty)$ *are quasinearly subharmonic if and only if there are constants* $K = K(u, \Omega, N) \geq 1$ *such that*

$$u(x) \leq \frac{K}{(f(r))^N} \int_{B^N(x,r)} u(y)\, dm_N(y)$$

for all closed balls $\overline{B^N(x,r)} \subset \Omega$.

(ii) *For all open sets* $\Omega \subseteq \mathbb{R}^N$ *and all Lebesgue measurable sets* $\Gamma \subseteq \Omega$ *the characteristic functions* χ_Γ *are quasinearly subharmonic if and only if there are constants* $K = K(\Gamma, \Omega, N)$ *such that the inequality*

$$(f(r))^N \leq K m_N(B^N(x,r) \cap \Gamma)$$

holds for all closed balls $\overline{B^N(x,r)} \subset \Omega$ *with* $x \in \Gamma$.

(iii) *There are a set* $A \subseteq (0, +\infty)$ *and a constant* $c > 1$ *such that:*

(iii$_1$) *The inequality* $f(r) \leq cr$ *holds for all* $r \in (0, +\infty)$.

(iii$_2$) $\ln A$ *is an* ε*-net in* $\mathbb{R}$ *for some* $\varepsilon > 0$.

(iii$_3$) *The inequality*

$$\frac{1}{c} r \leq f(r)$$

holds for all $r \in A$.

Note that condition (iii) of Theorem 2.2 holds if and only if the set $\ln(A) := \{\ln x : x \in A\}$ is an ε-net in $\mathbb{R}$ for some $\varepsilon > 0$. A characterization in terms of porosity for the sets A which are favorable for *bounded* open sets $\Omega \subseteq \mathbb{R}^N$ is proved in Theorem 2.18 below.

2.2. **Generalized mean value inequalities.** Inequality (2.1), characteristic for quasinearly subharmonic functions, can be generalized by certain distinct ways. Our first theorem characterizes nonnegative quasinearly subharmonic functions via mean values over some sets more general than just balls. For this purpose we recall similarities of the Euclidean space.

Let Ω and D be subsets of $\mathbb{R}^N$ with marked points $p_\Omega \in \Omega$ and $p_D \in D$. In what follows we always suppose that $Int\, D \neq \emptyset$ and $p_D \in Int\, D$. Denote by $Sim(p_D, p_\Omega)$ the set of all similarities $h : \mathbb{R}^N \to \mathbb{R}^N$ such that $h(p_D) = p_\Omega$ and $h(D) \subseteq \Omega$. Recall that h is a similarity if there is a positive number $k = k(h)$, the similarity constant of h, such that

$$|h(x) - h(y)| = k|x - y|$$

for all $x, y \in \mathbb{R}^N$. The group of all similarities of the Euclidean space $\mathbb{R}^N$ is sometimes denoted as $SM(\mathbb{R}^N)$, see e.g. [24], 5.1.14, and we also adopt this designation. Observe that each similarity $h \in SM(\mathbb{R}^N)$ can be written in the form

$$h(x) = k(h)Tx + a, \quad x \in \mathbb{R}^N,$$

where $k(h) > 0$, and $T : \mathbb{R}^N \to \mathbb{R}^N$ is an orthogonal linear mapping and $a \in \mathbb{R}^N$.

Theorem 2.4. ([31], Theorem 2.2, p. 352) *Let Ω be an open set in $\mathbb{R}^N$, $N \geq 2$, let D be a bounded, Lebesgue measurable set with the marked point $p_D \in Int\, D$ and let $u : \Omega \to [0, +\infty)$ be a function from $\mathcal{L}^1_{\text{loc}}(\Omega)$. Then u is quasinearly subharmonic if and only if there is $C \geq 1$ such that*

$$u(x_\Omega) \leq \frac{C}{m_N(h(D))} \int\limits_{h(D)} u(y)\, dm_N(y) \tag{2.5}$$

for every point x_Ω and all $h \in Sim(p_D, x_\Omega)$. If u is K-quasinearly subharmonic, then $C = C(D, p_D, K, N)$ and, conversely, if (2.5) holds, then u is K-quasinearly subharmonic with $K = K(D, p_D, C, N)$.

Proof. Write

$$R_D := \sup_{y \in D} |p_D - y|, \quad \text{and} \quad r_D := \delta_{Int(D)}(p_D). \tag{2.6}$$

Suppose that u is quasinearly subharmonic, i.e., there is $K \geq 1$ such that (2.1) holds for all $B^N(x, r) \subseteq \Omega$. Let x_Ω be an arbitrary point of Ω and let $h \in Sim(p_D, x_\Omega)$. The last membership relation implies the inclusions

$$B^N(x_\Omega, k(h) r_D) \subseteq \Omega \quad \text{and} \quad h(D) \subseteq \overline{B^N(x_\Omega, k(h) R_D)}$$

where $k(h)$ is the similarity constant of h. Consequently we obtain

$$\frac{1}{m_N(h(D))} \int\limits_{h(D)} u(y)\, dm_N(y) \geq \frac{1}{\nu_N (k(h) R_D)^N} \int\limits_{h(D)} u(y)\, dm_N(y) \geq$$
$$\geq \left(\frac{r_D}{R_D}\right)^N \frac{1}{\nu_N (k(h) r_D)^N} \int\limits_{B^N(x_\Omega, k(h) r_D)} u(y)\, dm_N(y) \geq \left(\frac{r_D}{R_D}\right)^N \frac{u(x_\Omega)}{K}.$$

Thus if f is K-quasinearly subharmonic, then (2.5) holds with

$$C = \frac{K (R_D)^N}{(r_D)^N}.$$

Conversely, suppose that (2.5) holds with some $C \geq 1$ for all x_Ω and all $h \in Sim(p_D, x_\Omega)$. Let $B^N(x_\Omega, r_0) \subseteq \Omega$. Let h be an arbitrary similarity with $k(h) = \frac{r_0}{R_D}$ and with $h(p_D) = x_\Omega$. Then we have $h \in Sim(p_D, x_\Omega)$ and $B^N(x_\Omega, k(h) r_D) \subseteq h(D) \subseteq \overline{B^N(x_\Omega, r_0)}$. Consequently

$$\frac{C}{m_N(B^N(x_\Omega, r_0))} \int\limits_{B^N(x_\Omega, r_0)} u(y)\, dm_N(y) \geq$$
$$\geq \frac{C m_N(h(D))}{m_N(B^N(x_\Omega, r_0)) m_N(h(D))} \int\limits_{h(D)} u(y)\, dm_N(y) \geq u(x_\Omega) \frac{m_N(h(D))}{m_N(B^N(x_\Omega, r_0))}.$$

Since

$$\frac{m_N(B^N(x_\Omega,r_0))}{m_N(h(D))}=\frac{\nu_N(r_0)^N}{m_N(h(D))}=\frac{\nu_N(k(h))^N(R_D)^N}{(k(h))^N m_N(D)}=\frac{\nu_N R_D^N}{m_N(D)},$$

inequality (2.2) holds with

$$K=C\frac{\nu_N(R_D)^N}{m_N(D)}.$$

□

Remark 2.5. The standard notion of quasinearly subharmonicity is defined by the condition (2.1), where $D=B^N(0,1)$, $p_D=0$ and the considered similarities h are of the form $h(x)=r_0x+x_\Omega$. The point of Theorem 2.4 is that the definition and its consequences are, however, much more general: Instead of just $D=B^N(0,1)$ and $p_D=0$ one may consider arbitrary bounded sets D with nonvoid interior $Int\, D$.

Remark 2.6. Inequality (2.5) remains valid for each nonnegative quasinearly subharmonic function if we use bi-Lipschitz mappings h instead of similarities, but in this more general case the constant in (2.5) depends on the Lipschitz constant of h. See Lemma 2.1 in [29].

Inequality (2.5) remains also valid for unbounded sets D if $m_N(D)<+\infty$.

Proposition 2.7. ([31], Proposition 2.3, p. 353) *Let Ω be an open set in $\mathbb{R}^N$, $N\geq 2$, D a Lebesgue measurable set with $m_N(D)<+\infty$, p_D a point of $Int(D)$ and let $u:\Omega\to[0,+\infty)$ be a K-quasinearly subharmonic function. Then there is a constant $C=C(D,p_D,K,N)$ such that (2.5) holds for all $x_\Omega\in\Omega$ and $h\in Sim(p_D,x_\Omega)$.*

Proof. If D is bounded, then this proposition follows from Theorem 2.5. Suppose D is unbounded. Let $t>1$ be a constant. It is easy to show that there is a ball $B^N(p_D,r_t)$ with a sufficiently large radius r_t such that

$$t\,m_N(D\cap B^N(p_D,r_t))\geq m_N(D). \tag{2.7}$$

Write

$$D_t:=D\cap B^N(p_D,r_t)\quad\text{and}\quad p_{D_t}:=p_D.$$

Note that D_t satisfies all the conditions of Theorem 2.5 and that $p_{D_t}\in Int(D_t)$. Consequently there is $K\geq 1$ such that the inequality

$$u(x_\Omega)\leq\frac{K}{m_N(h(D_t))}\int\limits_{h(D_t)}u(y)\,dm_N(y)$$

holds for all x_Ω and $h\in Sim(p_{D_t},x_\Omega)$. If $h\in Sim(p_D,x_\Omega)$, then we have $h\in Sim(p_{D_t},x_\Omega)$ and $h(D_t)\subseteq h(D)$. Since

$$\frac{m_N(D_t)}{m_N(D)}=\frac{m_N(h(D_t))}{m_N(h(D))}$$

for all $h\in SM(\mathbb{R}^N)$, (2.7) implies the inequality

$$\frac{1}{m_N(h(D_t))}\int\limits_{h(D_t)}u(y)\,dm_N(y)\leq\frac{t}{m_N(h(D))}\int\limits_{h(D)}u(y)\,dm_N(y).$$

Thus (2.5) holds for all $h \in Sim(p_D, x_\Omega)$ with $C = tK$. □

Remark 2.8. If $Sim(p_D, x_\Omega) = \emptyset$ for all x_Ω, then Proposition 2.8 is vacuously true.

Let $\varphi : SM(\mathbb{R}^N) \to (0, +\infty)$ be a function such that the equality

$$\varphi(h) = \varphi(is \circ h)$$

holds for all $h \in SM(\mathbb{R}^N)$ and for all isometries $is : \mathbb{R}^N \to \mathbb{R}^N$. Then we have $\varphi(h_1) = \varphi(h_2)$ whenever $k(h_1) = k(h_2)$, that is there is a function $f : (0, +\infty) \to (0, +\infty)$ such that the equality

$$\varphi(h) = f(k(h)) \tag{2.8}$$

is fulfilled for all $h \in SM(\mathbb{R}^N)$ with $k(h)$ equals the similarity constant of h. For instance, if D is a bounded nonvoid subset of $\mathbb{R}^N$, we can put $\varphi(h) = diam(h(D))$. Other examples can be found below in Examples 5 – 8, say.

Let D be a measurable subset of $\mathbb{R}^N$ with a marked point $p_D \in Int\,D$. For every open set $\Omega \subseteq \mathbb{R}^N$ define a subset $Q(f, D, \Omega) \subseteq \mathcal{L}^1_{loc}(\Omega)$ by the rule:

$u \in Q(f, D, \Omega)$ if and only if $u \geq 0$ and $u \in \mathcal{L}^1_{loc}(\Omega)$ and there is $K = K(u) \geq 1$ such that the inequality

$$u(x_\Omega) \leq \frac{K}{(f(k(h)))^N} \int_{h(D)} u(y)\, dm_N(y) \tag{2.9}$$

holds for every $x_\Omega \in \Omega$ and all $h \in Sim(p_D, x_\Omega)$.

It is clear that $QNS(\Omega) = Q(f, D, \Omega)$ if D satisfies the conditions of Theorem 2.5 and $f(k(h)) = k(h)(m_N(D))^{\frac{1}{n}}$.

Proposition 2.9. ([31], Proposition 2.4, p. 355) *Let D be a bounded, Lebesgue measurable subset of $\mathbb{R}^N$ with a marked point $p_D \in Int\,D$ and let $\varphi : SM(\mathbb{R}^N) \to (0, +\infty)$, $f : (0, +\infty) \to (0, +\infty)$ be functions such that (2.8) holds for all $h \in SM(\mathbb{R}^N)$. Then the inclusion*

$$QNS(\Omega) \subseteq Q(f, D, \Omega) \tag{2.10}$$

is valid for all open sets $\Omega \subseteq \mathbb{R}^N$ if and only if there is $c \geq 1$ such that the inequality

$$f(k) \leq ck \tag{2.11}$$

holds for all $k \in (0, +\infty)$.

Proof. Suppose that inclusion (2.10) holds for all open sets $\Omega \subseteq \mathbb{R}^N$. Let Ω be an open half-space of $\mathbb{R}^N$. Then for every $k_0 \in (0, +\infty)$ there is a similarity h_0 with the similarity constant $k(h_0) = k_0$ such that $h_0(D) \subseteq \Omega$. The constant function u_1, $u_1(x) \equiv 1$ for $x \in \Omega$, belongs to $QNS(\Omega)$. Hence, by (2.10), $u_1 \in Q(f, D, \Omega)$ and it follows from (2.9) that

$$1 = u_1(h_0(p_D)) \leq \frac{K}{(f(k_0))^N} \int_{h_0(D)} u_1(x)\, dm_N(x) = \frac{K m_N(h_0(D))}{(f(k_0))^N} = \frac{K (k_0)^N m_N(D)}{(f(k_0))^N}.$$

Consequently (2.10) implies (2.11) for all $k \in (0,+\infty)$ with

$$c = (K(u_1)m_N(D))^{\frac{1}{N}} \vee 1.$$

Conversely suppose that (2.11) holds for all $k \in (0,+\infty)$. Then using Theorem 2.5 we obtain the following inequalities for every open set $\Omega \subseteq \mathbb{R}^N$, every $u \in QNS(\Omega)$, every $x_\Omega \in \Omega$ and every $h \in Sim(p_D, x_\Omega)$:

$$\begin{aligned} u(x_\Omega) &\leq \frac{C(u)}{m_N(h(D))} \int\limits_{h(D)} u(x)\,dm_N(x) = \\ &= \frac{C(u)(f(k(h)))^N}{(k(h))^N m_N(D)(f(k(h)))^N} \int\limits_{h(D)} u(x)\,dm_N(x) \leq \\ &\leq \frac{C(u)c^N}{m_N(D)(f(k(h)))^N} \int\limits_{h(D)} u(x)\,dm_N(x). \end{aligned}$$

Hence (2.9) holds with

$$K = \frac{C(u)c^N}{m_N(D)} \vee 1.$$

Thus (2.10) is valid for all open sets $\Omega \subseteq \mathbb{R}^N$. □

Before passing to the equality

$$Q(f,D,\Omega) = QNS(\Omega)$$

we consider one relevant question.

Theorem 2.10. ([31], Theorem 2.5, pp. 355-356) *Let A be a subset of $(0,+\infty)$. Then A is favorable for all open sets $\Omega \subseteq \mathbb{R}^N$ if and only if the following statement holds.*

(s) *There exists $C = C(A) > 1$ such that*

$$\left[\frac{x}{C}, x\right] \cap A \neq \emptyset \tag{2.12}$$

for every $x \in (0,+\infty)$.

The following lemma will be used in the proof of Theorem 2.10.

Lemma 2.11. ([31], Lemma 2.6, p. 356) *Let $A \subseteq (0,+\infty)$. Statement (s) of Theorem 2.10 does not hold with this A, if and only if there are disjoint open intervals (a_m, b_m), $a_m < b_m$, $m = 1,2,\ldots$, in $(0,+\infty)\setminus A$ such that*

$$\lim_{m\to+\infty} \frac{a_m}{b_m} = 0 \tag{2.13}$$

and either

$$\lim_{m\to+\infty} a_m = \lim_{m\to+\infty} b_m = 0 \tag{2.14}$$

or

$$\lim_{m\to+\infty} a_m = \lim_{m\to+\infty} b_m = +\infty. \tag{2.15}$$

Proof. If statement (s) holds, then using (2.12) we obtain that

$$\frac{a}{b} \geq \frac{1}{C(A)}$$

for every open interval (a,b) in $(0,+\infty)\setminus A$. This inequality contradicts (2.13).

Conversely, suppose that statement (s) of Theorem 2.10 does not hold and that 0 and $+\infty$ are limit points of A. Then for every natural $i \geq 2$ there is $x \in (0,+\infty)$ such that

$$\left(\frac{x}{i}, x\right) \cap A = \emptyset.$$

Let $\bar{A}$ be the closure of A in $(0,+\infty)$. Write (a_i, b_i) for the connected component of $(0,+\infty)\setminus \bar{A}$ which contains $(\frac{x}{i}, x)$. Since both 0 and $+\infty$ are limit points of A we have

$$0 < a_i < b_i < +\infty.$$

Passing to convergent, in $[0,+\infty]$, subsequences $\{a_{i_m}\}_{m\in\mathbb{N}}$ and $\{b_{i_m}\}_{m\in\mathbb{N}}$ it is easy to see that limits $\lim_{m\to+\infty} a_{i_m}$ and $\lim_{m\to+\infty} b_{i_m}$ are 0 or $+\infty$ and that the equalities

$$\lim_{m\to+\infty} a_{i_m} = 0 \quad \text{and} \quad \lim_{m\to+\infty} b_{i_m} = +\infty$$

cannot be true simultaneously. Renaming $a_m := a_{i_m}$ and $b_m := b_{i_m}$ we obtain the desirable sequence of intervals in $(0,+\infty)\setminus A$.

If at least one of the points 0 and $+\infty$ is not a limit point of A, then there is $\varepsilon > 0$ such that

$$A \subset (0,\varepsilon] \quad \text{or} \quad A \subset [\varepsilon, +\infty).$$

Each of these inclusions implies evidently the existence of desired intervals in $(0,+\infty)\setminus \bar{A}$. □

Proof of Theorem 2.10. We shall first prove that A is favorable for all open sets $\Omega \subseteq \mathbb{R}^N$ if statement (s) holds.

Suppose that (s) is true. Let Ω be an open set in $\mathbb{R}^N$ and let $u \in \mathcal{L}^1_{\mathrm{loc}}(\Omega)$ be a nonnegative function which satisfies condition (ii) of Definition 2.1. It is enough to show that $u \in QNS(\Omega)$. To prove this, consider an arbitrary $\overline{B^N(x_\Omega, r_0)} \subseteq \Omega$. By statement (s) there is $r_1 \in A$ such that

$$\frac{r_0}{C} \leq r_1 \leq r_0$$

where the constant $C = C(A) > 1$. Using this double inequality and condition (ii) of Definition 2.1 we obtain

$$u(x_\Omega) \leq \frac{K}{\nu_N (r_1)^N} \int\limits_{B^N(x_\Omega, r_1)} u(y)\, dm_N(y) \leq \frac{KC^N}{\nu_N (r_0)^N} \int\limits_{B^N(x_\Omega, r_0)} u(y)\, dm_N(y).$$

Statement (iv) of Proposition 1.3 implies that $u \in QNS(\Omega)$.

Conversely, suppose that A is favorable for every open set $\Omega \subseteq \mathbb{R}^N$. We must show that (s) holds. If (s) does not hold then, by Lemma 2.11 there is a sequence of disjoint open intervals in $(0,+\infty)$ satisfying (2.13) and (2.14) or (2.13) and (2.15). Suppose that (2.13) and (2.14) hold. Then for every integer $N_0 > 2$ there is a sequence of open intervals (a_m, b_m) such that

$$0 < b_{m+1} < a_m < 2a_m < \frac{1}{N_0} b_m < b_m \tag{2.16}$$

and

$$(a_m, b_m) \cap A = \emptyset \tag{2.17}$$

for $m = 1, 2, \ldots$ and

$$\lim_{m \to +\infty} \frac{b_m}{a_m} = +\infty. \tag{2.18}$$

Moreover, passing, if necessary, to a subsequence we may assume that

$$\sum_{m=1}^{+\infty} b_m < +\infty. \tag{2.19}$$

For the sake of simplicity, we shall describe our constructions only on the plane but in such a way that a generalization to the dimensions $n \geq 3$ is a trivial matter.

Define the points $z_m \in \mathbb{C}$, $m = 1, 2, \ldots$, as

$$z_m := \begin{cases} 0, & \text{if } m = 1, \\ 2\sum_{i=1}^{m-1} b_i, & \text{if } m \geq 2 \end{cases}$$

and write

$$R_1 := \{z \in \mathbb{C} : 0 < Re(z) < 2b_1,\ |Im(z)| < a_2\}$$

and

$$R_m := \{z \in \mathbb{C} : 2\sum_{i=1}^{m-1} b_i < Re(z) < 2\sum_{i=1}^{m} b_i,\ |Im(z)| < a_{m+1}\}$$

for $m \geq 2$. Using (2.16) we see that $B^2(z_m, b_m)$ are open, pairwise disjoint balls and that R_m are open, pairwise disjoint rectangles. The desired domain Ω is, by definition, the union

$$\bigcup_{m=1}^{+\infty} \left(B^2\left(z_m, \frac{b_m}{N_0}\right) \cup R_m \right).$$

Let us define now a function u as the characteristic function of the set

$$X := \overline{\bigcup_{m=1}^{+\infty} B^2(z_m, a_m)}, \tag{2.20}$$

i.e.,

$$u(z) := \begin{cases} 1, & \text{if } z \in X, \\ 0, & \text{if } z \in \Omega \setminus X. \end{cases} \tag{2.21}$$

It is clear that $u \geq 0$ and that $u \in \mathcal{L}^1(\Omega)$. Moreover, since

$$\frac{1}{m_2(B^2(z_m, \frac{b_m}{2N_0}))} \int\limits_{B^2(z_m, \frac{b_m}{2N_0})} u(z)\, dm_2(z) = \frac{4N_0^2(a_m)^2}{(b_m)^2}, \tag{2.22}$$

statement (iv) of Proposition 1.3 and limit relation (2.18) imply $u \notin QNS(\Omega)$. It remains to show that there is K such that (2.3) holds whenever $r \in A$ and $\overline{B^2(x,r)} \subseteq \Omega$. If $x \in \Omega \setminus X$, then (2.3) is trivial and we must consider only $x \in X$. The last membership relation implies that there exists $m = m_x$ such that

$$x \in \overline{B^2(z_m, a_m)}. \tag{2.23}$$

Let us consider all $r \in A$ such that

$$\overline{B^2(x,r)} \subseteq \Omega. \tag{2.24}$$

From (2.17) follows that either $r \geq b_{m_x}$ or $r \leq a_{m_x}$. If $r \geq b_{m_x}$, then we have

$$B^2(x,r) \supseteq \overline{B^2(z_{m_x}, \frac{b_{m_x}}{N_0})}. \tag{2.25}$$

Indeed, the triangle inequality and (2.16) imply

$$|y - x| \leq |x - z_{m_x}| + |z_{m_x} - y| \leq a_{m_x} + \frac{b_{m_x}}{N_0} < \frac{3}{4} b_{m_x} < r$$

for all $y \in \overline{B^2(z_{m_x}, \frac{b_{m_x}}{N_0})}$. Inclusion (2.25) and the definition of Ω show that $B^2(x,r) \nsubseteq \Omega$ if $r \geq b_{m_x}$. Consequently, if (2.24) holds, then

$$r \leq a_{m_x}. \tag{2.26}$$

Using the last inequality, (2.23) and (2.16) we obtain

$$\overline{B^2(x,r)} \subseteq B^2\left(z_m, \frac{b_m}{N_0}\right), \quad m = m_x,$$

for these x and r. Hence the equality

$$\frac{1}{m_2(B^2(x,r))} \int\limits_{B^2(x,r)} u(y)\, dm_2(y) = \frac{m_2(B^2(z_m, a_m) \cap B^2(x,r))}{m_2(B^2(x,r))} \tag{2.27}$$

holds for such x and r. Write

$$C = \inf \frac{m_2(B^2(z_m, a_m) \cap B^2(x,r))}{m_2(B^2(x,r))} \tag{2.28}$$

where the infimum is taken over the set of all balls $B^2(x,r)$ with $x \in B^2(z_m, a_m)$ and with $r \leq a_m$. If r is fixed and $x_1, x_2 \in B^2(z_m, a_m)$, then the inequality $|x_1 - z_m| \geq |x_2 - z_m|$ implies

$$m_2(B^2(z_m, a_m) \cap B^2(x_1, r)) \leq m_2(B^2(z_m, a_m) \cap B^2(x_2, r)).$$

Thus we have

$$C = \inf_{r \leq a_m} \frac{m_2(B^2(z_m, a_m) \cap B^2(z_m + a_m, r))}{m_2(B^2(z_m + a_m, r))}.$$

The right-hand side of the last formula is invariant under the similarities. Consequently using the similarity

$$\mathbb{C} \ni z \longmapsto \frac{1}{r}(z - z_m) \in \mathbb{C}$$

we see that

$$(2.29) \quad C = \inf_{r \leq a_m} \frac{m_2(B^2(0, \frac{a_m}{r}) \cap B^2(\frac{a_m}{r}, 1))}{m_2(B^2(\frac{a_m}{r}, 1))} = \inf_{r \geq 1} \frac{m_2(B^2(0, r) \cap B^2(r, 1))}{m_2(B^2(r, 1))} =$$

$$= \frac{1}{\pi} \inf_{r \geq 1} m_2(B^2(0, r) \cap B^2(r, 1)) = \frac{1}{\pi} m_2(B^2(-1, 1) \cap B^2(0, 1)) = \frac{2}{3} - \frac{\sqrt{3}}{2\pi}.$$

The last equality, (2.26) and (2.28) imply that

$$\frac{1}{(\frac{2}{3} - \frac{\sqrt{3}}{2\pi}) m_2(B^2(x, r))} \int_{B^2(x,r)} u_3(y)\, dm_2(y) \geq u_3(x)$$

whenever $r \in A$ and $\overline{B^2(x, r)} \subseteq \Omega$.

Thus the theorem is proved in the case where limit relations (2.13) and (2.14) hold. Similar constructions can be realized if (2.13) and (2.15) hold and we omit them here. □

Statement (s) of Theorem 2.11 has a useful reformulation. For $A \subseteq (0, +\infty)$ define

$$\ln(A) := \{\ln x : x \in A\}$$

with $\ln(\emptyset) := \emptyset$. Then $\ln(A)$ is a subset of $\mathbb{R}$. Recall that a set $X \subseteq \mathbb{R}$ is an ε-net in $\mathbb{R}$, $\varepsilon > 0$, if

$$\mathbb{R} = \bigcup_{x \in X} B^1(x, \varepsilon).$$

Proposition 2.12. ([31], Proposition 2.7, p. 360) *Let A be a subset of* $(0, +\infty)$. *Then statement (s) in Theorem 2.10 is valid with this A if and only if there is* $\varepsilon > 0$ *such that* $\ln(A)$ *is an* ε*-net in* $\mathbb{R}$.

Proof. If (s) holds, then $\ln(A)$ is an ε-net with $\varepsilon = \ln C$ where C is the constant in (2.12). If (s) does not hold, then Lemma 2.11 implies that $\ln A$ is not an ε-net for any $\varepsilon > 0$. □

Using this proposition and analysing the first part of the proof of Theorem 2.11 we obtain the following

Proposition 2.13. ([31], Proposition 2.8, p. 360) *Let A be a subset of* $(0, +\infty)$. *The following three statements are equivalent.*

(i) *A is favorable for all domains of* $\mathbb{R}^N$.
(ii) *A is favorable for all open sets of* $\mathbb{R}^N$.

(iii) *There is* $\varepsilon > 0$ *such that* $\ln(A)$ *is an* ε*-net in* $\mathbb{R}$.

The condition for the set $A \subseteq (0,+\infty)$, to be favorable for all *bounded* domains Ω can be presented in terms of porosity of A, so recall the definition.

Definition 2.14. Let $A \subseteq (0,+\infty)$. The right hand porosity of A at zero is the quantity

$$p_0(A) := \limsup_{h\to 0+} \frac{l(h,A)}{h}$$

where $l(h,A)$ is the length of the longest interval in $[0,h] \setminus A$, $h > 0$.

Remark 2.15. It is easy to see that $0 \le p_0(A) \le 1$ for each $A \subseteq \mathbb{R}$. A variety of computations directly related to the notion of porosity can be found in [148], pp. 183–212.

Definition 2.16. Let $A \subseteq (0,+\infty)$. The right porosity index of A at 0, $i_0(A)$, is defined to be the supremum of all real numbers r for which there is a sequence of open intervals $\{(a_n,b_n)\}_{n\in\mathbb{N}}$, $a_n < b_n$, such that $\lim_{n\to\infty} a_n = \lim_{n\to\infty} b_n = 0$ and $(a_n,b_n) \subset (0,+\infty) \setminus A$ and

$$r < \frac{b_n - a_n}{a_n}$$

for each $n \in \mathbb{N}$.

If no such numbers r exist, then following the usual conversion we define $i_0(A) := 0$.

The following lemma is a particular case of Lemma $A_{2.13}$ from [148], p. 185.

Lemma 2.17. ([31], Lemma 2.11, p. 361) *The equality*

$$i_0(A) = \frac{p_0(A)}{1 - p_0(A)}$$

holds for each $A \subseteq (0,+\infty)$.

Theorem 2.18. ([31], Theorem 2.12, p. 361) *Let* A *be a subset of* $(0,+\infty)$. *Then* A *is favorable for all bounded domains* $\Omega \subseteq \mathbb{R}^N$ *if and only if* $p_0(A) < 1$.

Proof. It follows from Lemma 2.17 that $p_0(A) = 1$ if and only if $i_0(A) = 1$. Using the definition of porosity index $i_0(A)$ we can prove that the equality $i_0(A) = +\infty$ implies the existence of disjoint intervals $(a_n,b_n) \subset (0,+\infty) \setminus A$, $n = 1,2,\ldots$, such that equations (2.13) and (2.14) hold. It was shown in the proof of Theorem 2.10 that if (2.13) and (2.14) hold then there are a domain $\Omega \subseteq \mathbb{R}^N$ and a nonnegative $u \in \mathcal{L}^1_{\text{loc}}(\Omega) \setminus QNS(\Omega)$ such that (2.3) holds whenever $r \in A$ and $\overline{B^N(x,r)} \subseteq \Omega$. It remains to observe that inequality (2.19) implies $diam(\Omega) < +\infty$. Thus if A is favorable for all bounded domains $\Omega \subseteq \mathbb{R}^N$, then $p_0(A) < 1$.

Now note that if $p_0(A) < 1$, then the set $(-\infty,R) \cap \ln(A)$ is an ε-net, $\varepsilon = \varepsilon(R)$, in $(-\infty,R)$ for each $R \in \mathbb{R}$. Hence, reasoning as in the first part of the proof of Theorem 2.10 we can prove the implication

$$(p_0(A) < 1) \Rightarrow (A \text{ is favorable for every bounded domain } \Omega \subseteq \mathbb{R}^N).$$

□

Remark 2.19. As in Proposition 2.13 it is easy to prove that A is favorable for all bounded domains of $\mathbb{R}^N$ if and only if A is favorable for all bounded open subsets of $\mathbb{R}^N$. Theorem 2.18 remains valid even for unbounded domains and open sets $\Omega \subseteq \mathbb{R}^N$ if

$$\sup_{x\in\Omega}(\delta_\Omega(x)) < +\infty.$$

Remark 2.20. In complete analogy with Definition 2.16 we may define the quantity $i_\infty(A)$, the left porosity index of A at $+\infty$, after which Proposition 2.12 can be reformulated as:

Let $A \subset (0,+\infty)$. The following statements are equivalent:

(i) *A is favorable for all domains $\Omega \subseteq \mathbb{R}^N$.*

(ii) *The indexes $i_0(A)$ and $i_\infty(A)$ are less than infinity,*

$$i_0(A) \vee i_\infty(A) < +\infty.$$

(iii) *There is $\varepsilon > 0$ such that $\ln(A)$ is an ε-net in $\mathbb{R}^1$.*

Theorem 2.10, Proposition 2.12 and Theorem 2.18 imply the following.

Corollary 2.21. ([31], Corollary 2.13, p. 362) *Let A be a subset of $(0,+\infty)$ and let α, β be positive constants. Then the set A is favorable for all domains $\Omega \subseteq \mathbb{R}^N$ (for all bounded domains $\Omega \subseteq \mathbb{R}^N$) if and only if the set*

$$\alpha A^\beta := \{\alpha x^\beta : x \in A\}$$

has the same property.

Proof. One just directly observes that if condition (s) holds for the set A with a constant C, then condition (s) holds for the set αA^β with the constant $C' \geq C^\beta$. □

Now we are ready to characterize the function $f : (0,+\infty) \to (0,+\infty)$ for which the equality

$$Q(f,D,\Omega) = QNS(\Omega) \tag{2.30}$$

is fulfilled for all open sets $\Omega \subseteq \mathbb{R}^N$.

Theorem 2.22. ([31], Theorem 2.14, p. 362) *Let D be a bounded, Lebesgue measurable subset of $\mathbb{R}^N$ with a marked point $p_D \in Int\, D$ and let $\varphi : SM(\mathbb{R}^N) \to (0,+\infty)$ and $f : (0,+\infty) \to (0,+\infty)$ be functions such that (2.8) holds for all $h \in SM(\mathbb{R}^N)$. Then equality (2.9) holds for all open sets $\Omega \subseteq \mathbb{R}^N$ if and only if there are $A \subseteq (0,+\infty)$ and $c > 1$ such that:*

(i) *the inequality $f(k) \leq ck$ holds for all $k \in (0,+\infty)$,*

(ii) *$\ln(A)$ is an ε-net in $\mathbb{R}$ for some $\varepsilon > 0$,*

(iii) *the inequality*

$$\frac{1}{c}k \leq f(k) \tag{2.31}$$

holds for all $k \in A$.

Proof. Let Ω be an open set in $\mathbb{R}^N$, $A \subseteq (0,+\infty)$ and $c > 1$. Assume that $\ln(A)$ is an ε-net in $\mathbb{R}$ for some $\varepsilon > 0$ and that (2.8) holds with this c for all $k \in A$. Then using (2.9) and (2.31) we obtain

$$u(x_\Omega) \leq \frac{c^N K(u)}{(k(h))^N} \int_{h(D)} u(y)\, dm_N(y) \tag{2.32}$$

for every $u \in Q(f,D,\Omega)$ and every x_Ω whenever $h \in Sim(p_D, x_\Omega)$ and $k(h) \in A$. As in the proof of Theorem 2.2 write

$$R_D = \sup_{y \in D} |p_D - y|.$$

Let $B^N(x_\Omega, r_0)$ be a ball such that $r_0 = R_D k_0$, with $k_0 \in A$ and $\overline{B^N(x_\Omega, r_0)} \subset \Omega$. Then each similarity h such that $k(h) = k_0$ and $h(p_D) = x_\Omega$ belongs to $Sim(p_D, x_\Omega)$ and satisfies $h(D) \subseteq B^N(x_\Omega, r_0)$. Consequently (2.32) implies

$$\begin{aligned} u(x_\Omega) &\leq \frac{c^N K(u)(R_D)^N \nu_N}{(k_0 R_D)^N \nu_N} \int_{B^N(x_\Omega, r_0)} u(y)\, dm_N(y) = \\ &\leq \frac{c^N K(u) \nu_N (R_D)^N}{m_N(B^N(x_\Omega, r_0))} \int_{B^N(x_\Omega, r_0)} u(y)\, dm_N(y) \end{aligned} \tag{2.33}$$

for every $u \in Q(f,D,\Omega)$ and every $B^N(x_\Omega, r_0) \subseteq \Omega$ whenever $r_0 \in R_D A$. Corollary 2.21 implies that the set $R_D A$ is favorable for Ω. Hence $Q(f,D,\Omega) \subseteq QNS(\Omega)$. Taking into account Theorem 2.4 we see that conditions (i)–(iii) of the present theorem imply equality (2.30) for all open sets $\Omega \subseteq \mathbb{R}^N$.

Conversely, suppose that (2.30) holds for all open sets $\Omega \subseteq \mathbb{R}^N$ but for every $t > 0$ the set $\ln A_t$, where

$$A_t := \{k \in (0,+\infty) : f(k) \geq tk\}, \tag{2.34}$$

is not an ε-net for any $\varepsilon > 0$. It is clear that $A_{t_1} \subseteq A_{t_2}$ if $t_1 \geq t_2$. Applying Proposition 2.12 and Lemma 2.11 to the sets $A_2, A_3, \ldots$ we obtain a sequence $\{a_m\}_{m=2}^{+\infty}$ of positive numbers a_m such that $A_m \cap (a_m, m a_m) = \emptyset$ for each $m \geq 2$, i.e.,

$$f(k) < \frac{1}{m} k \tag{2.35}$$

if $a_m < k < m a_m$, and that

$$(a_{m_1}, m_1 a_{m_1}) \cap (a_{m_2}, m_2 a_{m_2}) = \emptyset \tag{2.36}$$

whenever $m_1 \neq m_2$. Passing, if necessary, to a subsequence we may assume that $\{a_m\}_{m=2}^{+\infty}$ and $\{m a_m\}_{m=2}^{+\infty}$ are monotone and convergent in $[0,+\infty]$ sequences. This assumption and (2.36) imply either the equalities

$$\lim_{m \to +\infty} a_m = \lim_{m \to +\infty} m a_m = 0 \tag{2.37}$$

or the equalities

$$\lim_{m\to+\infty} a_m = \lim_{m\to+\infty} ma_m = +\infty.$$

As in the proof of Theorem 2.10 we consider only the case when (2.37) holds and the dimension $n=2$. We shall construct a domain $\Omega\subseteq\mathbb{R}^2$ and a nonnegative $u\in\mathcal{L}^1_{\rm loc}(\Omega)$ such that

$$u\in Q(f,D,\Omega)\setminus QNS(\Omega).$$

To this end note that (2.30) implies (2.10), so using Proposition 2.4 we can find $c\geq 1$ such that

$$f(k)\leq ck \tag{2.38}$$

for every $k\in(0,+\infty)$. Let us define a function $f_1:(0,+\infty)\to(0,+\infty)$ by the rule

$$f_1(k) := \begin{cases} \frac{k}{m} & \text{if } a_m<k<ma_m,\ m=2,3,\dots \\ ck & \text{if } k\in(0,+\infty)\setminus\bigcup_{m=2}^{+\infty}(a_m,ma_m)\end{cases} \tag{2.39}$$

where $c\geq 1$ is the constant from inequality (2.38). Inequalities (2.35) and (2.38) imply $f(k)\leq f_1(k)$ for all $k\in(0,+\infty)$. Hence from the definition of the set $Q(f,D,\Omega)$ follows the inclusion

$$Q(f,D,\Omega)\supseteq Q(f_1,D,\Omega).$$

Thus it is sufficient to find a domain $\Omega\subseteq\mathbb{R}^2$ and a nonnegative $u\in\mathcal{L}^1_{\rm loc}(\Omega)$ such that

$$u\in Q(f_1,D,\Omega)\setminus QNS(\Omega).$$

Let us define

$$\Omega := \bigcup_{m=N_0+1}^{+\infty}\left(B^2\left(z_m,\frac{ma_m}{N_0}\right)\cup R_m\right),\quad u(x):=\begin{cases}1 & \text{if } x\in X\\ 0 & \text{if } x\in\Omega\setminus X\end{cases}$$

where

$$X := \bigcup_{m=N_0+1}^{+\infty}\overline{B^2(z_m,a_m)},\quad z_m := 2\sum_{i=1}^{m-1} ia_i$$

$$R_m := \{z\in\mathbb{C}: 2\sum_{n=1}^{m-1} na_n < Re(z) < 2\sum_{n=1}^{m} na_n,\quad |Im(z)|<a_{m+1}\}.$$

The parameter N_0 is free here and we will specify this parameter later. It is relevant to remark that the domain Ω is obtained from a domain by deleting of the balls $B^2(z_1,\frac{b_1}{N_0}), B^2(z_2,\frac{b_2}{N_0}),\dots,B^2(z_{N_0},\frac{b_{N_0}}{N_0})$ and the rectangels $R_1,R_2,\dots,R_{N_0}$ and putting $b_m:=ma_m$ in the rest of balls and rectangels. As in the proof of Theorem 2.11 we have $u\notin QNS(\Omega)$. It still remains to prove that $u\in Q(f_1,D,\Omega)$. The last relation holds if and only if there exists $K(u)\geq 1$ such that

$$(f_1(k(h)))^2\leq K(u)\int_{h(D)} u(y)\,dm_2(y) \tag{2.40}$$

for all $h\in Sim(p_D,x_\Omega)$ with $x_\Omega\in X$.

Let $x_\Omega \in X$. It follows from the definitions of Ω and X that there is $m \geq N_0 + 1$ for which

$$x_\Omega \in \overline{B^2(z_m, a_m)}.$$

We claim that the inequality

$$k(h) r_D \leq \frac{2ma_m}{N_0}. \tag{2.41}$$

holds for every $h \in Sim(p_D, x_\Omega)$ with $r_D = \delta_{Int(D)}(p_D)$.

Let us prove it. Since $h \in Sim(p_D, x_\Omega)$ we have

$$h(B^2(p_D, r_D)) \subseteq \Omega.$$

The last inclusion implies

$$\partial\Omega \cap h(B^2(p_D, r_D)) = \emptyset$$

because $\Omega \cap \partial\Omega = \emptyset$ for the open sets. The intersection

$$\partial B^2\left(z_m, \frac{ma_m}{N_0}\right) \cap \partial\Omega = \left\{z \in \mathbb{C} : |z - z_m| = \frac{ma_m}{N_0}\right\} \cap \partial\Omega$$

is not empty. Consequently there is $\xi \in \partial B^2(z_m, \frac{ma_m}{N_0}) \setminus h(B^2(p_D, r_D))$. Hence

$$|x_\Omega - \xi| \geq k(h) r_D.$$

Using the triangle inequality we obtain

$$|x_\Omega - \xi| \leq |x_\Omega - z_m| + |z_m - \xi| = |x_\Omega - z_m| + \frac{ma_m}{N_0}.$$

Consequently

$$k(h) r_D \leq |x_\Omega - z_m| + \frac{ma_m}{N_0}.$$

Since $x_\Omega \in \overline{B^2(z_m, a_m)}$ we have $|x_\Omega - z_m| \leq a_m$. It follows directly from the definition of Ω that $m \geq N_0$. Hence

$$k(h) r_D \leq a_m + \frac{ma_m}{N_0} \leq \frac{2ma_m}{N_0}.$$

Inequality (2.41) follows.

Since $h(D) \supseteq h(B^2(p_D, r_D))$, the inequality

$$(f_1(k(h)))^2 \leq K(u) \int\limits_{B^2(x_\Omega, k(h) r_D)} u(y)\, dm_2(y) \tag{2.42}$$

implies (2.40), so it is sufficient to prove (2.42). The following two cases are possible: $k(h) \in (0, a_m]$ and $k(h) \in (a_m, +\infty)$. Before analyzing these cases note that $f_1(k) \leq ck$ for every $k \in (0, +\infty)$ because $\frac{1}{m} \leq \frac{1}{2}$ and $c \geq 1$ in definition (2.39). Hence in the first case we can replace (2.42) by

$$c^2 \leq \frac{K(u)}{(k(h))^2} \int\limits_{B^2(x_\Omega, k(h) r_D)} u(y)\, dm_2(y). \tag{2.43}$$

It is clear that
(2.44)
$$\frac{K(u)}{(k(h))^2}\int_{B^2(x_\Omega,k(h)r_D)} u(y)\,dm_2(y) \geq \frac{K(u)\pi(r_D\wedge 1)^2}{\pi(k(h)(r_D\wedge 1))^2}\int_{B^2(x_\Omega,k(h)(r_D\wedge 1))} u(y)\,dm_2(y).$$

Since $k(h)\in(0,a_m]$, we see that
$$k(h)(r_D\wedge 1)\leq a_m.$$

Hence, as it was shown in the proof of Theorem 2.10, in the case under consideration we have
$$\frac{1}{\pi(k(h)(r_D\wedge 1))^2}\int_{B^2(x_\Omega,k(h)(r_D)\wedge 1)} u(y)\,dm_2(y)\geq \frac{2}{3}-\frac{\sqrt{3}}{2\pi}.$$

The last estimation and (2.44) show that (2.43) holds if
$$c^2 = K(u)\pi(r_D\wedge 1)^2\Big(\frac{2}{3}-\frac{\sqrt{3}}{2\pi}\Big).$$

Consider now the case $k(h)\in(a_m,+\infty)$. Inequality (2.41) shows that
$$k(h)\leq\frac{2ma_m}{N_0 r_D}.$$

Let us specify N_0 as the smallest positive integer N satisfying the inequality $\frac{2}{Nr_D}>1$. Then we obtain the double inequality
$$a_m<k(h)<ma_m.$$

This inequality and (2.39) show that
$$f_1(k(h))=\frac{k}{m}\leq a_m.$$

Consequently we can prove (2.42) as in the case $k(h)\in(0,a_m]$. □

Let us consider now some examples of functions φ and f for which equality (2.8) holds.

Example 5. Let ψ be a positive bounded periodic function on $\mathbb{R}$. Write
$$\mu(x):=\frac{1}{2}\Big(x+\frac{1}{x}\Big)$$
for $x>0$ and define

$$f(k):=k\psi(\mu(k)). \tag{2.45}$$

Using some routine estimations we see that conditions (i)–(iii) from Theorem 2.22 are satisfied by the function f if we take
$$A=\mu^{-1}\{x\in(0,+\infty):\psi(x)\geq\frac{1}{2}M\},\quad c=M\vee\frac{2}{M}$$

where

$$M = \sup_{y \in \mathbb{R}} \psi(y).$$

An important special case of the preceding example is the constant function ψ. Then f is linear on $(0,+\infty)$ and conditions (i)–(iii) from Theorem 2.22 evidently hold. In this simplest case the function $\varphi : SM(\mathbb{R}^n) \to (0,+\infty)$ connected with f can be obtained in distinct ways depending on the geometrical properties of the set D.

In all following examples D is a bounded Lebesgue measurable subset of $\mathbb{R}^N$ with $Int\, D \neq \emptyset$ and $h \in SM(\mathbb{R}^N)$.

Example 6. Let d-dimensional Hausdorff measure $\mathcal{H}^d$, $N-1 \leq d \leq N$, of the boundary ∂D be finite and nonzero, $0 < \mathcal{H}^d(\partial D) < +\infty$. Write

$$\varphi(h) = (\mathcal{H}^d(\partial(h(D))))^{\frac{1}{d}}.$$

Example 7. Let D be a set with the finite Cacciopoli–de Giorgi perimeter P, see, for instance, [12], Chapter 3. Write

$$\varphi(h) = (P(h(D)))^{\frac{1}{N-1}}.$$

Example 8. Let $D \subseteq \mathbb{R}^2$ be a simply connected domain with rectifiable boundary ∂D, $0 < \mathcal{H}^1(\partial D) < +\infty$. Suppose that the domain D is not a disk. Write

$$\varphi(h) = ((\mathcal{H}^1(h(\partial D)))^2 - 4\pi m_2(h(D)))^{\frac{1}{2}}.$$

In this case the inequality $\varphi(h) > 0$ follows from the Classical Isoperimetric Inequality, see, for instance, [12], Chapter 1.

This list of examples can be simply extended by involving the analytic capacity, the transfinite diameter, the Menger curvature etc. for the definition of the function φ. The homogenity under dilatations $x \longmapsto \alpha x$, $x \in \mathbb{R}^N$, $\alpha > 0$, the invariance under isometries, finiteness and positiveness are sufficient for this purpose.

3. A MEAN VALUE TYPE INEQUALITY FOR QUASINEARLY SUBHARMONIC FUNCTIONS

Abstract. Basing our proof on an old argument of Domar, we generalize and improve Armitage's and Gardiner's previous subharmonic function inequality result. Our result is stated for quasinearly subharmonic functions, it is rather general and, at the same time, flexible. Indeed, with the aid of it we will in the next two sections improve both Domar's and our previous domination condition results of subharmonic functions, and also Armitage's and Gardiner's, and our results on the subharmonicity of separately subharmonic functions.

Keywords. Subharmonic, quasinearly subharmonic, mean value, weighted integrals, control functions

3.1. **Older mean value type inequalities for subharmonic functions.** In [110], Lemma 3.2, p. 5, we gave a rather general inequality type result which is related to the inequalities (3.1) and (3.2), at least partly, see also [115], Lemma 2.2, p. 6. These results have their origins in the previous considerations of Armitage and Gardiner [3], proof of Proposition 2, pp. 257-259. Observe however, and as already Armitage and Gardiner have pointed out, this inequality is based on an old argument of Domar [27], pp. 431–432.

3.2. **A mean value type inequality for quasinearly subharmonic functions.** As pointed out already above, our previous result [110], Lemma 3.2, p. 5, was a generalized version of Armitage's and Gardiner's argument [3], proof of Proposition 2, pp. 257-258, and as such, it was based on an old argument of Domar [27], Lemma 1, pp. 431-432 and 430. Now we will give another variant of this inequality type result, Theorem 3.4 below. We begin with two preliminary lemmas.

3.2.1. ***Preliminary lemmas.*** The following result and its proof is essentially due to Domar [27], Lemma 1, pp. 431-432 and 430. We state the result, however, in a more general form, at least seemingly. See also [3], pp. 258-259.

Lemma 3.1. ([110], Lemma 3.1, p. 4) *Let $K \geq 1$. Let $\phi : [0,+\infty) \to [0,+\infty)$ be an increasing (strictly or not) function for which there exist s_0, $s_1 \in \mathbb{N}$, $s_0 < s_1$, such that $\phi(s) > 0$ and*

$$2K\phi(s-s_0) \leq \phi(s)$$

for all $s \geq s_1$. Let $u : D \to [0,+\infty)$ be a K-quasinearly subharmonic function. Suppose that

$$u(x_j) \geq \phi(j)$$

for some $x_j \in D$, $j \geq s_1$. If

$$R_j \geq \left(\frac{2K}{\nu_N}\right)^{1/N} \left[\frac{\phi(j+1)}{\phi(j)} m_N(A_j)\right]^{1/N},$$

where

$$A_j := \{x \in D : \phi(j-s_0) \leq u(x) < \phi(j+1)\},$$

then either $B^N(x_j,R_j)\cap(\mathbb{R}^N\setminus D)\neq\emptyset$ *or there is* $x_{j+1}\in B^N(x_j,R_j)$ *such that*

$$u(x_{j+1})\geq\phi(j+1).$$

Proof. Choose

$$R_j\geq\left(\frac{2K}{\nu_N}\right)^{1/N}\left[\frac{\phi(j+1)}{\phi(j)}m_N(A_j)\right]^{1/N},$$

and suppose that $B^N(x_j,R_j)\subset D$. Suppose on the contrary that $u(x)<\phi(j+1)$ for all $x\in B^N(x_j,R_j)$. Using the generalized mean value inequality for u we see that

$$\begin{aligned}
&\phi(j)\leq u(x_j)\leq\frac{K}{\nu_N R_j^N}\int\limits_{B^N(x_j,R_j)}u(x)\,dm_N(x)=\\
&\leq\frac{K}{\nu_N R_j^N}\int\limits_{B^N(x_j,R_j)\cap A_j}u(x)\,dm_N(x)+\frac{K}{\nu_N R_j^N}\int\limits_{B^N(x_j,R_j)\setminus A_j}u(x)\,dm_N(x)<\\
&<\left[\frac{Km_N(B^N(x_j,R_j)\cap A_j)}{\nu_N R_j^N}\frac{\phi(j+1)}{\phi(j)}+\frac{Km_N(B^N(x_j,R_j)\setminus A_j)}{\nu_N R_j^N}\frac{\phi(j-s_0)}{\phi(j)}\right]\phi(j)<\\
&<\phi(j),
\end{aligned}$$

a contradiction. □

The next lemma is a slightly generalized version of Armitage's and Gardiner's result [3], Proposition 2, p. 257. The proof of our refinement is – as already pointed out – a rather straightforward modification of Armitage's and Gardiner's argument [3], proof of Proposition 2, pp. 257-259.

Lemma 3.2. ([110], Lemma 3.2, pp. 4-7) *Let* $K\geq 1$. *Let* $\varphi:[0,+\infty)\to[0,+\infty)$ *and* $\psi:[0,+\infty)\to[0,+\infty)$ *be increasing functions for which there exist* $s_0,s_1\in\mathbb{N}$, $s_0<s_1$, *such that*

(i) *the inverse functions* φ^{-1} *and* ψ^{-1} *are defined on* $[\min\{\varphi(s_1-s_0),\psi(s_1-s_0)\},+\infty)$,
(ii) $2K(\psi^{-1}\circ\varphi)(s-s_0)\leq(\psi^{-1}\circ\varphi)(s)$ *for all* $s\geq s_1$,
(iii) $\sum_{j=s_1+1}^{+\infty}\left[\frac{(\psi^{-1}\circ\varphi)(j+1)}{(\psi^{-1}\circ\varphi)(j)}\frac{1}{\varphi(j-s_0)}\right]^{1/(N-1)}<+\infty.$

Let $u:D\to[0,+\infty)$ *be a K-quasinearly subharmonic function. Let* $\tilde{s}_1\in\mathbb{N}$, $\tilde{s}_1\geq s_1$, *be arbitrary. Then for each* $x\in D$ *and* $r>0$ *such that* $\overline{B^N(x,r)}\subset D$ *either*

$$u(x)\leq(\psi^{-1}\circ\varphi)(\tilde{s}_1+1)$$

or

$$\Phi(u(x))\leq\frac{C}{r^N}\int\limits_{B^N(x,r)}\psi(u(y))\,dm_N(y),$$

where $C = C(N,K,s_0)$ *and* $\Phi : [s_2, +\infty) \to [0, +\infty)$,

$$\Phi(t) := \left(\sum_{j=j_0}^{+\infty} \left[\frac{(\psi^{-1} \circ \varphi)(j+1)}{(\psi^{-1} \circ \varphi)(j)} \frac{1}{\varphi(j - s_0)} \right]^{1/(N-1)} \right)^{1-N},$$

and $j_0 \in \{s_1+1, s_1+2, \ldots\}$ *is such that*

$$(\psi^{-1} \circ \varphi)(j_0) \leq t < (\psi^{-1} \circ \varphi)(j_0+1),$$

and $s_2 := (\psi^{-1} \circ \varphi)(s_1+1)$.

Proof. Take $x \in D$ and $r > 0$ arbitrarily such that $\overline{B^N(x,r)} \subset D$. We may suppose that $u(x) > (\psi^{-1} \circ \varphi)(\tilde{s}_1 + 1)$. Since φ and ψ are increasing and $(\psi^{-1} \circ \varphi)(s) \to +\infty$ as $s \to +\infty$, there is an integer $j_0 \geq \tilde{s}_1 + 1$ such that

$$(\psi^{-1} \circ \varphi)(j_0) \leq u(x) < (\psi^{-1} \circ \varphi)(j_0+1).$$

Write $x_{j_0} := x$, $D_0 := B^N(x_{j_0}, r)$, and for each $j \geq j_0$,

$$A_j := \{ y \in D_0 : (\psi^{-1} \circ \varphi)(j - s_0) \leq u(y) < (\psi^{-1} \circ \varphi)(j+1) \},$$

$$R_j := \left(\frac{2K}{\nu_N} \right)^{1/N} \left[\frac{(\psi^{-1} \circ \varphi)(j+1)}{(\psi^{-1} \circ \varphi)(j)} m_N(A_j) \right]^{1/N}.$$

If $B^N(x_{j_0}, R_{j_0}) \cap (\mathbb{R}^N \setminus D_0) \neq \emptyset$, then clearly

$$r < R_{j_0} \leq \sum_{k=j_0}^{+\infty} R_k.$$

On the other hand, if $B^N(x_{j_0}, R_{j_0}) \subset D_0$, then by Lemma 3.1 (where now

$$\phi(s) := \begin{cases} (\psi^{-1} \circ \varphi)(s), & \text{when } s \geq s_1 - s_0, \\ \frac{s}{s_1 - s_0} \phi(s_1 - s_0), & \text{when } 0 \leq s < s_1 - s_0, \end{cases}$$

say), there is $x_{j_0+1} \in B^N(x_{j_0}, R_{j_0})$ such that $u(x_{j_0+1}) \geq (\psi^{-1} \circ \varphi)(j_0+1)$. Suppose that for $k = j_0, j_0+1, \ldots, j$,

$$B^N(x_k, R_k) \subset D_0, \; x_{k+1} \in B^N(x_k, R_k),$$

this for $k = j_0, j_0+1, \ldots, j-1$, and $u(x_k) \geq (\psi^{-1} \circ \varphi)(k)$. By Lemma 3.1 there is then $x_{j+1} \in B^N(x_j, R_j)$ such that $u(x_{j+1}) \geq (\psi^{-1} \circ \varphi)(j+1)$. Since u is locally bounded above and $(\psi^{-1} \circ \varphi)(k) \to +\infty$ as $k \to +\infty$, we may suppose that $B^N(x_{j+1}, R_{j+1}) \cap (\mathbb{R}^N \setminus D_0) \neq \emptyset$. But then

$$r < \text{dist}(x_{j_0}, x_{j_0+1}) + \text{dist}(x_{j_0+1}, x_{j_0+2}) + \cdots + \text{dist}(x_j, x_{j+1}) + \text{dist}(x_{j+1}, \mathbb{R}^N \setminus D_0),$$

thus

$$r < R_{j_0} + R_{j_0+1} + \cdots + R_j + R_{j+1} \leq \sum_{k=j_0}^{+\infty} R_k. \tag{3.1}$$

Using, for $j = j_0 - s_0, j_0 + 1 - s_0, \ldots$, the notation

$$a_j := \{y \in D_0 : (\psi^{-1} \circ \varphi)(j) \leq u(y) < (\psi^{-1} \circ \varphi)(j+1)\},$$

we get from (3.1):

$$\begin{aligned}
r &< \sum_{k=j_0}^{+\infty} \left(\frac{2K}{\nu_N}\right)^{1/N} \left[\frac{(\psi^{-1}\circ\varphi)(k+1)}{(\psi^{-1}\circ\varphi)(k)} m_N(A_k)\right]^{1/N} = \\
&< \left(\frac{2K}{\nu_N}\right)^{1/N} \sum_{k=j_0}^{+\infty} \left(\left[\frac{(\psi^{-1}\circ\varphi)(k+1)}{(\psi^{-1}\circ\varphi)(k)} \frac{1}{\varphi(k-s_0)}\right]^{1/N} [\varphi(k-s_0) m_N(A_k)]^{1/N}\right) < \\
&< \left(\frac{2K}{\nu_N}\right)^{1/N} \left(\sum_{k=j_0}^{+\infty} \left[\frac{(\psi^{-1}\circ\varphi)(k+1)}{(\psi^{-1}\circ\varphi)(k)} \frac{1}{\varphi(k-s_0)}\right]^{1/(N-1)}\right)^{(N-1)/N} \times \\
&\qquad \times \left[\sum_{k=j_0}^{+\infty} \varphi(k-s_0) m_N(A_k)\right]^{1/N} < \\
&< \left(\frac{2K}{\nu_N}\right)^{1/N} \left(\sum_{k=j_0}^{+\infty} \left[\frac{(\psi^{-1}\circ\varphi)(k+1)}{(\psi^{-1}\circ\varphi)(k)} \frac{1}{\varphi(k-s_0)}\right]^{1/(N-1)}\right)^{(N-1)/N} \times \\
&\qquad \times \left[\sum_{k=j_0}^{+\infty} \int_{A_k} \psi(u(y)) dm_N(y)\right]^{1/N} < \\
&< \left(\frac{2K}{\nu_N}\right)^{1/N} \left(\sum_{k=j_0}^{+\infty} \left[\frac{(\psi^{-1}\circ\varphi)(k+1)}{(\psi^{-1}\circ\varphi)(k)} \frac{1}{\varphi(k-s_0)}\right]^{1/(N-1)}\right)^{(N-1)/N} \times \\
&\qquad \times \left(\sum_{k=j_0}^{+\infty} \left[\sum_{j=k-s_0}^{k} \int_{a_j} \psi(u(y)) dm_N(y)\right]\right)^{1/N} < \\
&< \left[\frac{2(s_0+1)K}{\nu_N}\right]^{1/N} \left(\sum_{k=j_0}^{+\infty} \left[\frac{(\psi^{-1}\circ\varphi)(k+1)}{(\psi^{-1}\circ\varphi)(k)} \frac{1}{\varphi(k-s_0)}\right]^{1/(N-1)}\right)^{(N-1)/N} \times \\
&\qquad \times \left[\int_{D_0} \psi(u(y)) dm_N(y)\right]^{1/N}.
\end{aligned}$$

Thus

$$\Phi(u(x)) \leq \frac{C}{r^N} \int_{D_0} \psi(u(y)) dm_N(y),$$

where $C = C(N, K, s_0)$ and $\Phi : [s_2, +\infty) \to [0, +\infty)$,

$$\Phi(t) := \left(\sum_{k=j_0}^{+\infty} \left[\frac{(\psi^{-1} \circ \varphi)(k+1)}{(\psi^{-1} \circ \varphi)(k)} \frac{1}{\varphi(k - s_0)} \right]^{1/(N-1)} \right)^{1-N},$$

where $j_0 \in \{s_1 + 1, s_1 + 2, \ldots\}$ is such that

$$(\psi^{-1} \circ \varphi)(j_0) \leq t < (\psi^{-1} \circ \varphi)(j_0 + 1),$$

and $s_2 = (\psi^{-1} \circ \varphi)(s_1 + 1)$.

The function Φ may be extended to the whole interval $[0, +\infty)$, for example as follows:

$$\Phi(t) := \begin{cases} \Phi(t), & \text{when } t \geq s_2, \\ \frac{t}{s_2} \Phi(s_2), & \text{when } 0 \leq t < s_2. \end{cases}$$

□

Remark 3.3. Write $s_3 := \max\{s_1 + 3, (\psi^{-1} \circ \varphi)(s_1 + 3)\}$, say. (We may suppose that s_3 is an integer.) Suppose, that in addition to the assumptions (i), (ii), (iii) of Lemma 3.2, also the following assumption is satisfied:

(iv) *the function*

$$[s_1 + 1, +\infty) \ni s \mapsto \frac{(\psi^{-1} \circ \varphi)(s+1)}{(\psi^{-1} \circ \varphi)(s)} \frac{1}{\varphi(s - s_0)} \in \mathbb{R}$$

is decreasing.

Then one can replace the function $\Phi \mid [s_3, +\infty)$ by the function $\Phi_1 \mid [s_3, +\infty)$, where $\Phi_1 = \Phi_1^{\varphi,\psi} : [0, +\infty) \to [0, +\infty)$,

$$\Phi_1^{\varphi,\psi}(t) := \begin{cases} \left(\int\limits_{(\varphi^{-1} \circ \psi)(t) - 2}^{+\infty} \left[\frac{(\psi^{-1} \circ \varphi)(s+1)}{(\psi^{-1} \circ \varphi)(s)} \frac{1}{\varphi(s - s_0)} \right]^{1/(N-1)} ds \right)^{1-N}, & \text{when } t \geq s_3, \\ \frac{t}{s_3} \Phi_1^{\varphi,\psi}(s_3), & \text{when } 0 \leq t < s_3. \end{cases}$$

Similarly, if the function

$$[s_1 + 1, +\infty) \ni s \mapsto \frac{(\psi^{-1} \circ \varphi)(s+1)}{(\psi^{-1} \circ \varphi)(s)} \in \mathbb{R}$$

is bounded, then in Lemma 3.2 one can replace the function $\Phi \mid [s_3, +\infty)$ by the function $\Phi_2 \mid [s_3, +\infty)$, where $\Phi_2 = \Phi_2^{\varphi,\psi} : [0, +\infty) \to [0, +\infty)$,

$$(3.2) \qquad \Phi_2^{\varphi,\psi}(t) := \begin{cases} \left[\int\limits_{(\varphi^{-1} \circ \psi)(t) - 2}^{+\infty} \frac{ds}{\varphi(s - s_0)^{1/(N-1)}} \right]^{1-N}, & \text{when } t \geq s_3, \\ \frac{t}{s_3} \Phi_2^{\varphi,\psi}(s_3), & \text{when } 0 \leq t < s_3. \end{cases}$$

3.2.2. *A mean value type inequality for quasinearly subharmonic functions.*

Theorem 3.4. ([116], Theorem 2.1, pp. 398-399) *Let $K \geq 1$. Let $\varphi: [0,+\infty] \to [0,+\infty]$ and $\psi: [0,+\infty] \to [0,+\infty]$ be increasing functions such that there are $s_0, s_1 \in \mathbb{N}$, $s_0 < s_1$, such that*

(i) *the inverse functions φ^{-1} and ψ^{-1} are defined on $[\min\{\varphi(s_1 - s_0), \psi(s_1 - s_0)\}, +\infty]$,*
(ii) *$2K(\psi^{-1} \circ \varphi)(s - s_0) \leq (\psi^{-1} \circ \varphi)(s)$ for all $s \geq s_1$,*
(iii) $\int\limits_{s_1}^{+\infty} \left[\frac{(\psi^{-1}\circ\varphi)(s+2)}{(\psi^{-1}\circ\varphi)(s)} \frac{1}{\varphi(s-s_0)}\right]^{\frac{1}{N-1}} ds < +\infty.$

Let $u : D \to [0,+\infty)$ be a K-quasinearly subharmonic function. Let $\tilde{s}_1 \in \mathbb{N}$, $\tilde{s}_1 \geq s_1$, be arbitrary. Then for each $x \in D$ and $r > 0$ such that $\overline{B^N(x,r)} \subset D$ either

$$u(x) \leq (\psi^{-1} \circ \varphi)(\tilde{s}_1 + 1)$$

or

$$\Phi(u(x)) \leq \frac{C}{r^N} \int\limits_{B^N(x,r)} \psi(u(y))\, dm_N(y),$$

where $C = C(N, K, s_0)$ and $\Phi : [0,+\infty) \to [0,+\infty)$,

$$\Phi(t) := \begin{cases} \left(\int\limits_{(\varphi^{-1}\circ\psi)(t)-2}^{+\infty} \left[\frac{(\psi^{-1}\circ\varphi)(s+2)}{(\psi^{-1}\circ\varphi)(s)} \cdot \frac{1}{\varphi(s-s_0)}\right]^{\frac{1}{N-1}} ds \right)^{1-N}, & \text{when } t \geq s_3, \\ \frac{t}{s_3}\Phi(s_3), & \text{when } 0 \leq t < s_3. \end{cases}$$

Here $s_3 := \max\{s_1 + 3, s_2, (\psi^{-1} \circ \varphi)(s_1 + 3)\}$ and $s_2 := \max\{s_1, (\psi^{-1} \circ \varphi)(s_1 + 1)\}$.

Proof. The proof follows at once from Lemma 3.2. As a matter of fact, it is sufficient to observe the following:

- Instead of the, in [110] used control functions $\varphi : [0,+\infty) \to [0,+\infty)$ and $\psi : [0,+\infty) \to [0,+\infty)$ one can equally well use the present control functions $\varphi : [0,+\infty] \to [0,+\infty]$ and $\psi : [0,+\infty] \to [0,+\infty]$. See [115], p. 5.
- If

  $$(\psi^{-1} \circ \varphi)(j_0) \leq t < (\psi^{-1} \circ \varphi)(j_0 + 1),$$

  then

$$\sum_{k=j_0}^{+\infty} \left[\frac{(\psi^{-1} \circ \varphi)(k+1)}{(\psi^{-1} \circ \varphi)(k)} \cdot \frac{1}{\varphi(k - s_0)}\right]^{\frac{1}{N-1}} \leq \tag{3.3}$$

$$\leq \int\limits_{(\varphi^{-1}\circ\psi)(t)-2}^{+\infty} \left[\frac{(\psi^{-1} \circ \varphi)(s+2)}{(\psi^{-1} \circ \varphi)(s)} \cdot \frac{1}{\varphi(s - s_0)}\right]^{\frac{1}{N-1}} ds.$$

□

In section 4 we will then apply the obtained inequality type result to domination conditions for families of quasinearly subharmonic functions, improving Domar's, Rippon's and our previous results, see [27], Theorem 1 and Theorem 2, pp. 430-431, [125], Theorem 5, p. 128, and [115], Theorem 2.1, pp. 4-5. In addition, in section 5 we apply this inequality to the quasinearly subharmonicity of separately quasinearly subharmonic functions, slightly improving our previous results [110], Theorem 4.1, pp. 8-9, and [113], Theorem 3.3.1, pp. e2621-e2622.

4. Domination Conditions for Families of Quasinearly Subharmonic Functions

Abstract. Using our version of a mean value type inequality for quasinearly subharmonic functions, presented in the previous section, we give domination conditions for families of quasinearly subharmonic functions. Our results improve previous results of Domar, Rippon and ours, and thus also the original results of Sjöberg and Brelot.

Keywords. Subharmonic, quasinearly subharmonic, family of quasinearly subharmonic functions, domination conditions

4.1. **Previous results.** We begin by recalling the results of Domar and Rippon. Let $F : D \to [0,+\infty]$ be an upper semicontinuous function. Let $\mathcal{F}$ be a family of subharmonic functions $u : D \to [0,+\infty)$, which satisfy the condition

$$u(x) \leq F(x) \text{ for all } x \in D.$$

Write

$$w(x) := \sup_{u \in \mathcal{F}} u(x),$$

and let $w^* : D \to [0,+\infty]$ be the upper semicontinuous regularization of w, that is,

$$w^*(x) := \limsup_{y \to x} w(y).$$

Improving the original results of Sjöberg [137], Théorème II, p. 314, and Brelot [11], Théorème 1, p. 6, cf. also Green [42], Theorem 2, p. 830, Domar [27], Theorem 1 and Theorem 2, pp. 430-431, gave the following result:

Theorem 4.1. *If for some* $\varepsilon > 0$,

$$\int_D [\log^+ F(x)]^{N-1+\varepsilon}\, dm_N(x) < +\infty,$$

then w is locally bounded above in D, and thus w^* is subharmonic in D.

As Domar points out, his method of proof applies also to more general functions, that is, to the above defined nonnegative quasinearly subharmonic functions. Much later Rippon [125], Theorem 1, p. 128, generalized Domar's result in the following form:

Theorem 4.2. *Let* $\varphi : [0,+\infty] \to [0,+\infty]$ *be an increasing function such that*

$$\int_1^{+\infty} \frac{dt}{\varphi(t)^{\frac{1}{N-1}}} < +\infty.$$

If

$$\int_D \varphi(\log^+ F(x))\, dm_N(x) < +\infty,$$

then w is locally bounded above in D, and thus w^* is subharmonic in D.

Juhani Riihentaus

As pointed out by Domar, [27], pp. 436-440, and by Rippon [125], p. 129, the above results are for many particular cases sharp.

4.2. **An improvement to the results of Domar and Rippon.** In [115], Theorem 2.1, pp. 4-5, we gave a general and at the same time flexible result which includes both Domar's and Rippon's results. Now we improve our result still further, see also [124], Theorem 2.1, pp. 129-130, and Theorem 2.8, p. 133.

Theorem 4.3. ([116], Theorem 3.3, p. 400) *Let* $K \geq 1$. *Let* $\varphi : [0,+\infty] \to [0,+\infty]$ *and* $\psi : [0,+\infty] \to [0,+\infty]$ *be increasing functions for which there are* $s_0, s_1 \in \mathbb{N}$, $s_0 < s_1$, *such that*

(i) *the inverse functions* φ^{-1} *and* ψ^{-1} *are defined on* $[\min\{\varphi(s_1 - s_0), \psi(s_1 - s_0)\}, +\infty]$,
(ii) $2K(\psi^{-1} \circ \varphi)(s - s_0) \leq (\psi^{-1} \circ \varphi)(s)$ *for all* $s \geq s_1$,
(iii) *the following integral is convergent:*

$$\int_{s_1}^{+\infty} \left[\frac{(\psi^{-1} \circ \varphi)(s+2)}{(\psi^{-1} \circ \varphi)(s)} \cdot \frac{1}{\varphi(s - s_0)} \right]^{\frac{1}{N-1}} ds < +\infty.$$

Let $\mathcal{F}_K$ *be a family of* K*-quasinearly subharmonic functions* $u : D \to [-\infty, +\infty)$ *such that*

$$u(x) \leq F_K(x) \; for \; all \; x \in D,$$

where $F_K : D \to [0,+\infty]$ *is a Lebesgue measurable function. If for each compact set* $E \subset D$,

$$\int_E \psi(F_K(x))\, dm_N(x) < +\infty,$$

then the family $\mathcal{F}_K$ *is locally (uniformly) bounded in* D. *Moreover, the function* $w^* : D \to [0,+\infty)$ *is* K*-quasinearly subharmonic. Here*

$$w^*(x) := \limsup_{y \to x} w(y),$$

where

$$w(x) := \sup_{u \in \mathcal{F}_K} u^+(x).$$

Proof. Let E be an arbitrary compact subset of D. Write $\rho_0 :=\text{dist}(E, \partial D)$. Clearly $\rho_0 > 0$. Write

$$E_1 := \bigcup_{x \in E} \overline{B^N(x, \frac{\rho_0}{2})}.$$

Then E_1 is compact, and $E \subset E_1 \subset D$. Take $u \in \mathcal{F}_K^+$ arbitrarily, where

$$\mathcal{F}_K^+ := \{u^+ : u \in \mathcal{F}_K\}.$$

Let $\tilde{s}_1 = s_1 + 2$, say. Take $x \in E$ arbitrarily and suppose that $u(x) > \tilde{s}_3$, where $\tilde{s}_3 := \max\{\, s_1 + 5, (\psi^{-1} \circ \varphi)(s_1 + 5)\,\}$, say. By Theorem 3.4 we have,

$$\Phi(u(x)) \leq \frac{C}{\left(\frac{\rho_0}{2}\right)^N} \int\limits_{B^N(x,\frac{\rho_0}{2})} \psi(u(y))\, dm_N(y) \leq \frac{C}{\left(\frac{\rho_0}{2}\right)^N} \int\limits_{E_1} \psi(F_K(y))\, dm_N(y) < +\infty,$$

where

$$\Phi(t) := \left(\int\limits_{(\varphi^{-1}\circ\psi)(t)-2}^{+\infty} \left[\frac{(\psi^{-1}\circ\varphi)(s+2)}{(\psi^{-1}\circ\varphi)(s)} \cdot \frac{1}{\varphi(s-s_0)} \right]^{\frac{1}{N-1}} ds \right)^{1-N}.$$

Now we know that

$$\int\limits_{s_1}^{+\infty} \left[\frac{(\psi^{-1}\circ\varphi)(s+2)}{(\psi^{-1}\circ\varphi)(s)} \cdot \frac{1}{\varphi(s-s_0)} \right]^{\frac{1}{N-1}} ds < +\infty.$$

Therefore, the set of function values

$$u(x), \quad x \in E, \quad u \in \mathcal{F}_K^+,$$

is bounded above.

The rest of the proof goes then as in [115], proof of Theorem 2.1, pp. 7-8. □

Remark 4.4. In Theorem 4.3 one can, instead of the assumption (iii), use also the following:

(iii') *the following series is convergent:*

$$\sum_{j=s_1+1}^{+\infty} \left[\frac{(\psi^{-1}\circ\varphi)(j+1)}{(\psi^{-1}\circ\varphi)(j)} \cdot \frac{1}{\varphi(j-s_0)} \right]^{\frac{1}{N-1}} < +\infty.$$

Instead of the above used function Φ one uses then the function $\Phi_1 : [s_2, +\infty) \to [0, +\infty)$,

$$\Phi_1(t) := \left(\sum_{k=j_0}^{+\infty} \left[\frac{(\psi^{-1}\circ\varphi)(k+1)}{(\psi^{-1}\circ\varphi)(k)} \cdot \frac{1}{\varphi(k-s_0)} \right]^{\frac{1}{N-1}} \right)^{1-N},$$

where $j_0 \in \{\, s_1, s_1 + 2, \ldots\}$ is such that

$$(\psi^{-1}\circ\varphi)(j_0) \leq t < (\psi^{-1}\circ\varphi)(j_0 + 1).$$

The function Φ_1 may of course be extended to the whole interval $[0, +\infty)$, for example as follows:

$$\Phi_1(t) := \begin{cases} \Phi_1(t), & \text{when } t \geq s_3, \\ \frac{t}{s_3}\Phi_1(s_3), & \text{when } 0 \leq t < s_3. \end{cases}$$

Before giving examples, we write Theorem 4.3 in the following, perhaps more concrete form, see also [115], Remark 2.5, p. 8.

Corollary 4.5. ([116], Corollary 3.5, p. 402) *Let $K \geq 1$. Let $\varphi : [0,+\infty] \to [0,+\infty]$ and $\phi : [0,+\infty] \to [0,+\infty]$ be strictly increasing surjections for which there are $s_0, s_1 \in \mathbb{N}$, $s_0 < s_1$, such that*

(i) $2K\phi^{-1}(e^{s-s_0}) \leq \phi^{-1}(e^s)$ *for all* $s \geq s_1$,

(ii) *the following integral is convergent:*

$$\int_{s_1}^{+\infty} \left[\frac{\phi^{-1}(e^{s+2})}{\phi^{-1}(e^s)} \frac{1}{\varphi(s-s_0)}\right]^{\frac{1}{N-1}} ds < +\infty.$$

Let $\mathfrak{F}_K$ be a family of K-quasinearly subharmonic functions $u : D \to [-\infty,+\infty)$ such that

$$u(x) \leq F_K(x) \text{ for all } x \in D,$$

where $F_K : D \to [0,+\infty]$ is a Lebesgue measurable function. If for each compact set $E \subset D$,

$$\int_E \varphi(\log^+ \phi(F_K(x)))\, dm_N(x) < +\infty,$$

then the family $\mathfrak{F}_K$ is locally (uniformly) bounded in D. Moreover, the function $w^* : D \to [0,+\infty)$ is K-quasinearly subharmonic. Here

$$w^*(x) := \limsup_{y \to x} w(y),$$

where

$$w(x) := \sup_{u \in \mathfrak{F}_K} u^+(x).$$

Proof. For the proof, just choose $\psi(t) = \varphi(\log^+ \phi(t))$. Since only big values count, we may simply use the formula $\psi(t) = \varphi(\log \phi(t))$. One sees easily that for some $\tilde{s}_2 \geq s_1$,

$$(\psi^{-1} \circ \varphi)(s) = \phi^{-1}(e^s) \text{ for all } s \geq \tilde{s}_2.$$

It is then easy to see that the assumptions of Theorem 4.3 are satisfied. □

Example 9. Let $\varphi : [0,+\infty] \to [0,+\infty]$ be a strictly increasing surjection such that (for some $s_0 \in \mathbb{N}$),

$$\int_{s_1}^{+\infty} \frac{ds}{\varphi(s-s_0)^{\frac{1}{N-1}}} < +\infty.$$

Choosing then various functions ϕ which, together with φ, satisfy the conditions (i) and (ii) of Corollary 4.5, one gets more concrete results. If ϕ and ϕ^{-1} satisfy (at least far away) the Δ_2-condition, then the conditions (i) and (ii) are surely satisfied (see also [115], Remark 2.5, p. 8). Typical choices for ϕ might be, say, the following:

$$\phi(t) := \frac{t^p}{(\log t)^q}, \; p > 0, q \in \mathbb{R}.$$

The choice $p = 1$, $q = 0$ gives then the results of Domar and Rippon, Theorem 4.1 and Theorem 4.2 above. Choosing $0 < p < 1$ and $q \geq 0$, one gets (at least formal) improvements.

Observe that here, and below, we consider increasing functions $\phi : [0, +\infty] \to [0, +\infty]$, say, and which are of a certain form far away, that is, for big values of the argument. In such a case, we take the liberty to use the convention that the function is then automatically defined for small values of the argument in an approriate way. As an example, when we write, as above, for $p > 0$, $q \in \mathbb{R}$, say,

$$\phi(t) = \frac{t^p}{(\log t)^q},$$

we mean in fact the following function:

$$\phi(t) := \begin{cases} \frac{t^p}{(\log t)^q}, & \text{when } t \geq t_1, \\ \frac{t}{t_1}\phi(t_1), & \text{when } 0 \leq t < t_1, \end{cases}$$

where $t_1 \geq 2$ is some suitable integer in $\mathbb{N}$, say. Observe that we will use the same convention and understanding in Example 10 and also below in Examples 11, 12 and 13.

Example 10. Let $\phi : [0, +\infty] \to [0, +\infty]$ be a strictly increasing surjection for which there is $t_1 > 1$ such that

$$s = \phi(t) = \log t \ \text{ for } \ t \geq t_1$$

and thus

$$t = \phi^{-1}(s) = e^s \ \text{ for } \ s \geq \log t_1.$$

One sees easily that the condition (i) of Corollary 4.5 holds. In this case ϕ^{-1} does not satisfy a Δ_2-condition. Therefore, in order to apply Corollary 4.5 to a family $\mathcal{F}_K$ of K-quasinearly subharmonic functions, we must choose an approriate strictly increasing surjection $\varphi : [0, +\infty] \to [0, +\infty]$ such that for some $\tilde{s}_1 \in \mathbb{N}$,

$$\int_{\tilde{s}_1}^{+\infty} \left[\frac{\phi^{-1}(e^{s+2})}{\phi^{-1}(e^s)} \cdot \frac{1}{\varphi(s - s_0)} \right]^{\frac{1}{N-1}} ds = \int_{\tilde{s}_1}^{+\infty} \frac{e^{\frac{e^2-1}{N-1}e^s}}{\varphi(s - s_0)^{\frac{1}{N-1}}} ds < +\infty.$$

Therefore, we have two restrictions for φ. As the first condition the above quite strong restriction and, as a second one, the following, at least seemingly a mild condition: For each compact set $E \subset D$,

$$\int_E \varphi(\log^+(\log F_K(x)))\, dm_N(x) < +\infty.$$

5. On the Subharmonicity of Separately Subharmonic Functions and Generalizations

Abstract. With the aid of our new version of a mean value type inequality, we improve our previous result of the subharmonicity of separately subharmonic functions, and thus also the well-known related earlier results of Armitage and Gardiner and ours. Moreover, we give refinements, with concise proofs, to the basic classical results of Avanissian, of Lelong, and of Arsove.

Keywords. Subharmonic, quasinearly subharmonic, separately quasinearly subharmonic function

5.1. **Subharmonic functions versus separately subharmonic functions.** It is well-known that subharmonic functions need not be separately subharmonic. A simple counter example is the function $v : \mathbb{C}^n \to [-\infty, +\infty)$

$$u(z_1, \dots, z_n) = -\frac{1}{|z|^2} = -\frac{1}{\sum_{i=1}^n |z_i|^2} \quad (n \geq 2).$$

On the other hand, the question whether a separately subharmonic function is subharmonic stayed open for a rather long time. This problem was solved by Wiegerinck, who showed that a separately subharmonic function need not be subharmonic, see [155], Theorem, p. 770, and Wiegerinck and Zeinstra [156], Theorem 1, p. 246. We recall here this important result of Wiegerinck, with its original proof.

Theorem 5.1. *There exist a nonnegative function f on $\mathbb{C}^2$ with the following properties.*

(i) *For every $z_0, w_0 \in \mathbb{C}$, $fz_0, \cdot)$ and $f(\cdot, w_0)$ are continuous subharmonic functions on $\mathbb{C}$.*
(ii) *The function f is not subharmonic on $\mathbb{C}^2$.*

Proof. Let $a_n = (1/n)e^{i/(n+1)}$, $n = 1, 2, \dots$. Put

$$K_n' = \{z \in \mathbb{C} : |z| \leq n, \, 1/n \leq \arg z \leq 2\pi\} \cup \{0\},$$
$$K_n = K_n' \cup \{a_n\}, \; n = 1, 2, \dots.$$

By Runge's theorem the holomorphic function

$$f_n(z) := 0 \text{ on a small neighborhood of } K_n',$$
$$f_n(z) := n + 1 \text{ on a small neighborhood of } a_n,$$

can be uniformly approximated by polynomials on K_n. Hence there exist polynomials P_n such that $|P_n| < 1/2$ on K_n', $P_n(a_n) = n + 1$. Now let $h_n(z) = \max\{|P_n(z)| - 1, 0\}$, a continuous subharmonic function on $\mathbb{C}$, which equals 0 on K_n'.

The required function will be

$$f(z, w) = \sum_{n=1}^{+\infty} [h_n(z) h_n(w)].$$

Juhani Riihentaus

Observe that for fixed $z \in \mathbb{C}$ only finitely many of the $h_n(z)$ are nonzero. Therefore $f(z,\cdot)$ is a sum of finitely many continuous subharmonic functions and thus continuous and subharmonic. The same is true for $f(\cdot,w)$. However $f(a_n,a_n)$ tends to $+\infty$ with n, so we see that f is not locally bounded from above and hence not subharmonic on $\mathbb{C}^2$. □

Remark 5.2. It is easy to see that the function f is measurable and in fact continuous on $D := \mathbb{C}^2 \setminus \{(z,w) : \operatorname{Im} z = \operatorname{Im} w = 0,\ \operatorname{Re} z \geq 0,\ \operatorname{Re} w \geq 0\}$. Adapting the construction slightly, one can obtain that f becomes smooth in each variable separately as well as on D. In the definition of f one just uses instead of h_n the convolution $h_n(z - a_n/n) \star \phi_n$, where ϕ_n is a smooth, positive, radial function with sufficiently small support, while $\int \phi_n = 1$.

Remark 5.3. In [49] the question was asked for hyperharmonic functions, a slightly more general concept than superharmonic functions. Nevertheless, $-f$ gives a negative answer in this setting, because it is not even lower semicontinuous.

Remark 5.4. Generalizing a result of Cegrell and Sadullaev [14], Theorem 3.2, p. 83, Kalantar has recently shown that a separately subharmonic function is subharmonic outside a closed nowhere dense set with no bounded components, see [56], Theorem 3.1.

5.2. **Sufficient conditions for a separately subharmonic function to be subharmonic.** As pointed out above, separately subharmonic functions need not be subharmonic. On the other hand, imposing some integrability or related conditions on a separately subharmonic function, the subharmonicity follows, see e.g. [116]] and [117] and the detailed list of references therein. As a matter of fact, we consider this question below in quite detail.

5.2.1. ***Armitage's and Gardiner's result.*** Armitage and Gardiner [3], Theorem 1, p. 256, showed that a separately subharmonic function u in a domain Ω of $\mathbb{R}^{m+n}$, $m \geq n \geq 2$, is subharmonic, provided $\phi \circ \log^+ u^+ \in \mathcal{L}^1_{\mathrm{loc}}(\Omega)$, where $\phi : [0,+\infty) \to [0,+\infty)$ is an increasing function such that

$$\int_1^{+\infty} \frac{s^{\frac{n-1}{m-1}}}{\phi(s)^{\frac{1}{m-1}}} ds < +\infty.$$

Armitage's and Gardiner's result included all the previous existing results, that is, the results of Lelong [65], Théorème 1 bis, p. 315, of Avanissian [6], Théorème 9, p. 140, see also [67], Proposition 3, p. 24, and [49], Theorem, p. 31, of Arsove [5], Theorem 1, p. 622, and of ours [94], Theorem 1, p. 69. See also [14], Theorem 1.1, p. 79. Though Armitage's and Gardiner's result was close to being sharp, see [3], p. 255, it was, however, possible to improve their result slightly further. This was done in [110], Theorem 4.1, pp. 8-9, with the aid of quasinearly subharmonic functions. See also [113], Theorem 3.3.1 and Corollary 3.3.3, pp. e2621-e2622.

5.2.2. ***A further improvement.*** Now we improve our previous improvement to Armitage's and Gardiner's result still further.

The following standard notation will be used below: If Ω is a domain in $\mathbb{R}^{m+n}$, $m,n \geq 2$, and if $x \in \mathbb{R}^m$, $y \in \mathbb{R}^n$, then we write

$$\Omega(y) := \{x' \in \mathbb{R}^m : (x',y) \in \Omega\}, \quad \Omega(x) := \{y' \in \mathbb{R}^n : (x,y') \in \Omega\}.$$

Theorem 5.5. ([116], Theorem 4.1, p. 404) *Let $K \geq 1$. Let Ω be a domain in $\mathbb{R}^{m+n}$, $m \geq n \geq 2$. Let $u : \Omega \to [-\infty,+\infty)$ be a Lebesgue measurable function. Suppose that the following conditions are satisfied:*

(a) *For each $y \in \mathbb{R}^n$ the function*

$$\Omega(y) \ni x \mapsto u(x,y) \in [-\infty,+\infty)$$

is K-quasinearly subharmonic.

(b) *For each $x \in \mathbb{R}^m$ the function*

$$\Omega(x) \ni y \mapsto u(x,y) \in [-\infty,+\infty)$$

is K-quasinearly subharmonic.

(c) *There are increasing functions $\varphi : [0,+\infty) \to [0,+\infty)$ and $\psi : [0,+\infty) \to [0,+\infty)$ and s_0, $s_1 \in \mathbb{N}$, $s_0 < s_1$, such that*

(c1) *the inverse functions φ^{-1} and ψ^{-1} are defined on $[\min\{\varphi(s_1-s_0),\psi(s_1-s_0)\},+\infty)$,*

(c2) $2K(\psi^{-1}\circ\varphi)(s-s_0) \leq (\psi^{-1}\circ\varphi)(s)$ *for all $s \geq s_1$,*

(c3) *the following integral is convergent:*

$$\int\limits_{s_5}^{+\infty} \left[\frac{(\psi^{-1}\circ\varphi)(s+2)}{(\psi^{-1}\circ\varphi)(s)} \cdot \frac{1}{\varphi(s-s_0)}\right]^{\frac{1}{m-1}} \cdot \left(\int\limits_{s_5+s_0+2}^{s+s_0+2} \left[\frac{(\psi^{-1}\circ\varphi)(t+2)}{(\psi^{-1}\circ\varphi)(t)}\right]^{\frac{1}{n-1}} dt \right)^{\frac{n-1}{m-1}} ds < +\infty,$$

where $s_5 := \max\{s_0+s_1+3, s_0+(\psi^{-1}\circ\varphi)(s_1+3), s_0+(\varphi^{-1}\circ\psi)(s_1+3)\}$,

(c4) $\psi\circ u^+ \in \mathcal{L}^1_{\mathrm{loc}}(\Omega)$.

Then u is quasinearly subharmonic in Ω.

Proof. Recall that $s_2 = \max\{s_1,(\psi^{-1}\circ\varphi)(s_1+1)\}$ and $s_3 = \max\{s_1+3,s_2,(\psi^{-1}\circ\varphi)(s_1+3)\}$. Write $s_4 := \max\{s_0+s_3,(\varphi^{-1}\circ\psi)(s_1+3)\}$, say. Clearly, $s_0 < s_1 < s_3 < s_4 < s_5$. (We may of course suppose that s_3, s_4 and s_5 are integers.) One may replace u by $\max\{u^+,M\}$, where $M = \max\{s_5+3,(\psi^{-1}\circ\varphi)(s_4+3),(\varphi^{-1}\circ\psi)(s_4+3)\}$, say. We continue to denote u_M by u.

Step 1 *Use of Theorem 3.4.*

Take $(x_0,y_0) \in \Omega$ and $r > 0$ arbitrarily such that $\overline{B^m(x_0,2r)\times B^n(y_0,2r)} \subset \Omega$. Take $(\xi,\eta) \in B^m(x_0,r)\times B^n(y_0,r)$ arbitrarily. We know that $u(\cdot,y)$ is K-quasinearly subharmonic for each $y \in B^n(y_0,2r)$. In order to apply Theorem 3.4, it is clearly sufficient to show that

$$\int_{s_5+s_0+2}^{+\infty} \left[\frac{(\psi^{-1}\circ\varphi)(s+2)}{(\psi^{-1}\circ\varphi)(s)}\cdot\frac{1}{\varphi(s-s_0)}\right]^{\frac{1}{m-1}} ds < +\infty.$$

But this follows at once from the assumption (c3), since for all $s \geq s_5+s_0+2$,

$$\left(\int_{s_5+s_0+2}^{s+s_0+2} \left[\frac{(\psi^{-1}\circ\varphi)(t+2)}{(\psi^{-1}\circ\varphi)(t)}\right]^{\frac{1}{n-1}} dt\right)^{\frac{n-1}{m-1}} \geq \left(\int_{s_5+s_0+2}^{s+s_0+2} 1dt\right)^{\frac{n-1}{m-1}} = (s-s_5)^{\frac{n-1}{m-1}} \geq (s_0+2)^{\frac{n-1}{m-1}}.$$

From Theorem 3.4 it then follows that for all $y \in B^n(y_0, 2r)$

$$\tilde{\Phi}(u(\xi,y)) \leq \frac{C}{r^m} \int_{B^m(\xi,r)} \psi(u(x,y))dm_m(x), \tag{5.1}$$

where

$$\tilde{\Phi}(t) := \begin{cases} \left(\int_{(\varphi^{-1}\circ\psi)(t)-2}^{+\infty} \left[\frac{(\psi^{-1}\circ\varphi)(s+2)}{(\psi^{-1}\circ\varphi)(s)}\frac{1}{\varphi(s-s_0)}\right]^{\frac{1}{m-1}} ds\right)^{1-m}, & \text{when } t \geq s_3, \\ \frac{t}{s_3}\tilde{\Phi}(s_3), & \text{when } 0 \leq t < s_3. \end{cases}$$

Step 2 *Take mean values on both sides of* (5.1).

Taking (generalized) mean values with respect to the variable y over the ball $B^n(\eta, r)$ on both sides of (5.1), we get:

$$\begin{aligned} \frac{C}{r^n}\int_{B^n(\eta,r)} \tilde{\Phi}(u(\xi,y))dm_n(y) &\leq \frac{C}{r^n}\int_{B^n(\eta,r)} [\frac{C}{r^m}\int_{B^m(\xi,r)} \psi(u(x,y))dm_m(x)]\, dm_n(y) \\ &\leq \frac{C}{r^{m+n}} \int_{B^m(\xi,r)\times B^n(\eta,r)} \psi(u(x,y))dm_{m+n}(x,y) \\ &\leq \frac{C}{r^{m+n}} \int_{B^m(x_0,2r)\times B^n(y_0,2r)} \psi(u(x,y))dm_{m+n}(x,y). \end{aligned}$$

Here one must of course check that both $\psi\circ u(\cdot,\cdot)$ and $\tilde{\Phi}(u(\xi,\cdot))$ are Lebesgue measurable!

Step 3 *In order to apply Theorem 3.4 once more, define new functions* φ_1 *and* ψ_1.

Write $\psi_1 : [0,+\infty) \to [0,+\infty)$,

$$\psi_1(t) := \tilde{\Phi}(t) = \begin{cases} \left(\int_{(\varphi^{-1}\circ\psi)(t)-2}^{+\infty} \left[\frac{(\psi^{-1}\circ\varphi)(s+2)}{(\psi^{-1}\circ\varphi)(s)}\frac{1}{\varphi(s-s_0)}\right]^{\frac{1}{m-1}} ds\right)^{1-m}, & \text{when } t \geq s_3, \\ \frac{t}{s_3}\tilde{\Phi}(s_3), & \text{when } 0 \leq t < s_3. \end{cases}$$

It is easy to see that ψ_1 is defined, strictly increasing and continuous. Write then $\varphi_1 : [0,+\infty) \to [0,+\infty)$,

$$\varphi_1(t) := \begin{cases} \psi_1((\psi^{-1}\circ\varphi)(t)) = \tilde{\Phi}(\psi^{-1}(\varphi(t))), & \text{when } t \geq s_3, \\ \frac{t}{s_3}\psi_1((\psi^{-1}\circ\varphi)(s_3)) = \frac{t}{s_3}\tilde{\Phi}(\psi^{-1}(\varphi(s_3))), & \text{when } 0 \leq t < s_3. \end{cases}$$

Also φ_1 is defined, strictly increasing and continuous. This follows from the facts that ψ_1 is defined, strictly increasing and continuous (similarly as the functions $\varphi|[s_1 - s_0,+\infty)$ and $\psi|[s_1 - s_0,+\infty)$). Observe here that for $t \geq s_4$, say,

$$\varphi_1(t) = \left(\int\limits_{(\varphi^{-1}\circ\psi)((\psi^{-1}\circ\varphi)(t))-2}^{+\infty} \left[\frac{(\psi^{-1}\circ\varphi)(s+2)}{(\psi^{-1}\circ\varphi)(s)} \frac{1}{\varphi(s-s_0)} \right]^{\frac{1}{m-1}} ds \right)^{1-m} =$$
$$= \left(\int\limits_{t-2}^{+\infty} \left[\frac{(\psi^{-1}\circ\varphi)(s+2)}{(\psi^{-1}\circ\varphi)(s)} \frac{1}{\varphi(s-s_0)} \right]^{\frac{1}{m-1}} ds \right)^{1-m}.$$

One sees easily that $(\psi_1^{-1}\circ\varphi_1)(t) = (\psi^{-1}\circ\varphi)(t)$ for all $t \geq s_3$, thus $2K(\psi_1^{-1}\circ\varphi_1)(s - s_0) \leq (\psi_1^{-1}\circ\varphi_1)(s)$ for all $s \geq \bar{s}_1 \geq s_4$.

To show that

$$\int\limits_{s_5+s_0+2}^{+\infty} \left[\frac{(\psi_1^{-1}\circ\varphi_1)(s+2)}{(\psi_1^{-1}\circ\varphi_1)(s)} \frac{1}{\varphi_1(s-s_0)} \right]^{\frac{1}{n-1}} ds < +\infty,$$

we proceed as follows.

Write $F : [s_5,+\infty) \times [s_5+s_0+2,+\infty) \to [0,+\infty)$,

$$F(s,t) := \begin{cases} \left[\frac{(\psi_1^{-1}\circ\varphi_1)(t+2)}{(\psi_1^{-1}\circ\varphi_1)(t)} \frac{(\psi^{-1}\circ\varphi)(s+2)}{(\psi^{-1}\circ\varphi)(s)} \frac{1}{\varphi(s-s_0)} \right]^{\frac{1}{m-1}}, & \text{when } s_5+s_0+2 \leq t-s_0-2 \leq s, \\ 0, & \text{when } s_5 \leq s < t-s_0-2. \end{cases}$$

Suppose that $m > n \geq 2$. Then just calculate, use Minkowski's inequality and assumption (c3):

$$\left(\int_{s_5+s_0+2}^{+\infty}\left[\frac{(\psi_1^{-1}\circ\varphi_1)(t+2)}{(\psi_1^{-1}\circ\varphi_1)(t)}\frac{1}{\varphi_1(t-s_0)}\right]^{\frac{1}{n-1}}dt\right)^{\frac{n-1}{m-1}} =$$

$$= \left(\int_{s_5+s_0+2}^{+\infty}\left[\frac{(\psi_1^{-1}\circ\varphi_1)(t+2)}{(\psi_1^{-1}\circ\varphi_1)(t)}\right]^{\frac{1}{n-1}}\left(\int_{t-s_0-2}^{+\infty}\left[\frac{(\psi^{-1}\circ\varphi)(s+2)}{(\psi^{-1}\circ\varphi)(s)}\frac{1}{\varphi(s-s_0)}\right]^{\frac{1}{m-1}}ds\right)^{-\frac{1-m}{n-1}}dt\right)^{\frac{n-1}{m-1}} =$$

$$= \left(\int_{s_5+s_0+2}^{+\infty}\left[\frac{(\psi_1^{-1}\circ\varphi_1)(t+2)}{(\psi_1^{-1}\circ\varphi_1)(t)}\right]^{\frac{1}{n-1}}\left(\int_{t-s_0-2}^{+\infty}\left[\frac{(\psi^{-1}\circ\varphi)(s+2)}{(\psi^{-1}\circ\varphi)(s)}\frac{1}{\varphi(s-s_0)}\right]^{\frac{1}{m-1}}ds\right)^{\frac{m-1}{n-1}}dt\right)^{\frac{n-1}{m-1}} =$$

$$= \left(\int_{s_5+s_0+2}^{+\infty}\left(\int_{t-s_0-2}^{+\infty}\left(\left[\frac{(\psi_1^{-1}\circ\varphi_1)(t+2)}{(\psi_1^{-1}\circ\varphi_1)(t)}\cdot\frac{(\psi^{-1}\circ\varphi)(s+2)}{(\psi^{-1}\circ\varphi)(s)}\frac{1}{\varphi(s-s_0)}\right]^{\frac{1}{m-1}}\right)ds\right)^{\frac{m-1}{n-1}}dt\right)^{\frac{n-1}{m-1}} =$$

$$= \left(\int_{s_5+s_0+2}^{+\infty}\left[\int_{s_5}^{+\infty}F(s,t)\,ds\right]^{\frac{m-1}{n-1}}dt\right)^{\frac{n-1}{m-1}} \leq \left(\int_{s_5}^{+\infty}\left[\int_{s_5+s_0+2}^{+\infty}F(s,t)^{\frac{m-1}{n-1}}\,dt\right]^{\frac{n-1}{m-1}}ds\right) =$$

$$= \int_{s_5}^{+\infty}\left(\int_{s_5+s_0+2}^{s+s_0+2}\left(\left[\frac{(\psi_1^{-1}\circ\varphi_1)(t+2)}{(\psi_1^{-1}\circ\varphi_1)(t)}\right]^{\frac{1}{m-1}}\left[\frac{(\psi^{-1}\circ\varphi)(s+2)}{(\psi^{-1}\circ\varphi)(s)}\frac{1}{\varphi(s-s_0)}\right]^{\frac{1}{m-1}}\right)^{\frac{m-1}{n-1}}dt\right)^{\frac{n-1}{m-1}}ds =$$

$$= \int_{s_5}^{+\infty}\left[\frac{(\psi^{-1}\circ\varphi)(s+2)}{(\psi^{-1}\circ\varphi)(s)}\frac{1}{\varphi(s-s_0)}\right]^{\frac{1}{m-1}}\left(\int_{s_5+s_0+2}^{s+s_0+2}\left[\frac{(\psi_1^{-1}\circ\varphi_1)(t+2)}{(\psi_1^{-1}\circ\varphi_1)(t)}\right]^{\frac{1}{n-1}}dt\right)^{\frac{n-1}{m-1}}ds =$$

$$= \int_{s_5}^{+\infty}\left[\frac{(\psi^{-1}\circ\varphi)(s+2)}{(\psi^{-1}\circ\varphi)(s)}\frac{1}{\varphi(s-s_0)}\right]^{\frac{1}{m-1}}\left(\int_{s_5+s_0+2}^{s+s_0+2}\left[\frac{(\psi^{-1}\circ\varphi)(t+2)}{(\psi^{-1}\circ\varphi)(t)}\right]^{\frac{1}{n-1}}dt\right)^{\frac{n-1}{m-1}}ds < +\infty.$$

The case $m = n$ is considered similarly, just replacing Minkowski's inequality with Fubini's theorem.

Step 4 *Apply Theorem 3.4 to conclude that $u(\cdot,\cdot)$ is bounded in $B^m(x_0,r)\times B^n(y_0,r)$.*

With the aid of Theorem 3.4 we get

$$\Psi(u(\xi,\eta)) \leq \frac{C}{r^n}\int_{B^n(\eta,r)}\tilde{\Phi}(u(\xi,y))dm_n(y) \leq \frac{C}{r^{m+n}}\int_{B^m(x_0,2r)\times B^n(y_0,2r)}\psi(u(x,y))dm_{m+n}(x,y),$$

where now

$$\Psi(t) := \begin{cases} \left(\int\limits_{(\varphi_1^{-1}\circ\psi_1)(t)-2}^{+\infty} \left[\frac{(\psi_1^{-1}\circ\varphi_1)(s+2)}{(\psi_1^{-1}\circ\varphi_1)(s)} \frac{1}{\varphi_1(s-s_0)} \right]^{\frac{1}{n-1}} ds \right)^{1-n}, & \text{when } t \geq s_3, \\ \frac{t}{s_3}\Psi(s_3), & \text{when } 0 \leq t < s_3, \end{cases}$$

or equivalently

$$\Psi(t) := \begin{cases} \left(\int\limits_{(\varphi^{-1}\circ\psi)(t)-2}^{+\infty} \left[\frac{(\psi_1^{-1}\circ\varphi_1)(s+2)}{(\psi_1^{-1}\circ\varphi_1)(s)} \frac{1}{\varphi_1(s-s_0)} \right]^{\frac{1}{n-1}} ds \right)^{1-n}, & \text{when } t \geq s_3, \\ \frac{t}{s_3}\Psi(s_3), & \text{when } 0 \leq t < s_3. \end{cases}$$

Observe that we know that

$$\int\limits_{s_5+s_0+2}^{+\infty} \left[\frac{(\psi_1^{-1}\circ\varphi_1)(s+2)}{(\psi_1^{-1}\circ\varphi_1)(s)} \frac{1}{\varphi_1(s-s_0)} \right]^{\frac{1}{n-1}} ds \leq$$

$$\leq \int\limits_{s_5}^{+\infty} \left[\frac{(\psi^{-1}\circ\varphi)(s+2)}{(\psi^{-1}\circ\varphi)(s)} \frac{1}{\varphi(s-s_0)} \right]^{\frac{1}{m-1}} \left(\int\limits_{s_5+s_0+2}^{s+s_0+2} \left[\frac{(\psi^{-1}\circ\varphi)(t+2)}{(\psi^{-1}\circ\varphi)(t)} \right]^{\frac{1}{n-1}} dt \right)^{\frac{n-1}{m-1}} ds,$$

and that by assumption (c3),

$$\int\limits_{s_5}^{+\infty} \left[\frac{(\psi^{-1}\circ\varphi)(s+2)}{(\psi^{-1}\circ\varphi)(s)} \frac{1}{\varphi(s-s_0)} \right]^{\frac{1}{m-1}} \left(\int\limits_{s_5+s_0+2}^{s+s_0+2} \left[\frac{(\psi^{-1}\circ\varphi)(t+2)}{(\psi^{-1}\circ\varphi)(t)} \right]^{\frac{1}{n-1}} dt \right)^{\frac{n-1}{m-1}} ds < +\infty.$$

Hence the set of function values

$$(\varphi_1^{-1}\circ\psi_1)(u(\xi,\eta)) - 2 = (\varphi^{-1}\circ\psi)(u(\xi,\eta)) - 2, \ (\xi,\eta) \in B^m(x_0,r)\times B^n(y_0,r),$$

must be bounded. Thus the function $u(\cdot,\cdot)$ is bounded above in $B^m(x_0,r)\times B^n(y_0,r)$. By [109], Proposition 3.1, p. 57 (or by [113], Proposition 3.2.1, p. e2620, see also Proposition 5.8 below), we see that $u(\cdot,\cdot)$ is quasinearly subharmonic. □

Corollary 5.6. ([116], Corollary 4.2, p. 409) *Let $K \geq 1$. Let Ω be a domain in $\mathbb{R}^{m+n}$, $m \geq n \geq 2$. Let $u : \Omega \to [-\infty, +\infty)$ be a Lebesgue measurable function. Suppose that the following conditions are satisfied:*

(a) *For each $y \in \mathbb{R}^n$ the function*

$$\Omega(y) \ni x \mapsto u(x,y) \in [-\infty, +\infty)$$

is K-quasinearly subharmonic.

(b) *For each $x \in \mathbb{R}^m$ the function*

$$\Omega(x) \ni y \mapsto u(x,y) \in [-\infty, +\infty)$$

is K-quasinearly subharmonic.

(c) *There are strictly increasing surjections* $\varphi : [0,+\infty) \to [0,+\infty)$ *and* $\phi : [0,+\infty) \to [0,+\infty)$ *and* $s_0, s_1 \in \mathbb{N}$, $s_0 < s_1$, *such that*
(c1) $2K\phi^{-1}(e^{s-s_0}) \leq \phi^{-1}(e^s)$ *for all* $s \geq s_1$,
(c2) *the following integral is convergent:*

$$\int_{s_5}^{+\infty} \left[\frac{\phi^{-1}(e^{s+2})}{\phi^{-1}(e^s)} \frac{1}{\varphi(s-s_0)} \right]^{\frac{1}{m-1}} \left(\int_{s_5+s_0+2}^{s+s_0+2} \left[\frac{\phi^{-1}(e^{t+2})}{\phi^{-1}(e^t)} \right]^{\frac{1}{n-1}} dt \right)^{\frac{n-1}{m-1}} ds < +\infty.$$

(c3) $\varphi \circ \log^+ \phi(u^+) \in \mathcal{L}^1_{\mathrm{loc}}(\Omega)$.

Then u is quasinearly subharmonic in Ω.

Corollary 5.7. ([116], Corollary 4.3, pp. 409–410) *Let* $K \geq 1$. *Let* Ω *be a domain in* $\mathbb{R}^{m+n}$, $m \geq n \geq 2$. *Let* $u : \Omega \to [-\infty,+\infty)$ *be a Lebesgue measurable function. Suppose that the following conditions are satisfied:*

(a) *For each* $y \in \mathbb{R}^n$ *the function*

$$\Omega(y) \ni x \mapsto u(x,y) \in [-\infty,+\infty)$$

is K-quasinearly subharmonic.

(b) *For each* $x \in \mathbb{R}^m$ *the function*

$$\Omega(x) \ni y \mapsto u(x,y) \in [-\infty,+\infty)$$

is K-quasinearly subharmonic.

(c) *There are strictly increasing surjections* $\varphi : [0,+\infty) \to [0,+\infty)$ *and* $\phi : [0,+\infty) \to [0,+\infty)$ *and* $s_0, s_1 \in \mathbb{N}$, $s_0 < s_1$, *such that*
(c1) $2K\phi^{-1}(e^{s-s_0}) \leq \phi^{-1}(e^s)$ *for all* $s \geq s_1$,
(c2) ϕ^{-1} *satisfies a* Δ_2*-condition,*
(c3) *the following integral is convergent:*

$$\int_{s_1}^{+\infty} \frac{s^{\frac{n-1}{m-1}}}{\varphi(s-s_0)^{\frac{1}{m-1}}} ds < +\infty.$$

(c4) $\varphi \circ \log^+ \phi(u^+) \in \mathcal{L}^1_{\mathrm{loc}}(\Omega)$.

Then u is quasinearly subharmonic in Ω.

Example 11. Let u be separately subharmonic in Ω. Let $\varphi : [0,+\infty) \to [0,+\infty)$ be a strictly increasing surjection such that

$$\int_{s_1}^{+\infty} \frac{s^{\frac{n-1}{m-1}}}{\varphi(s-s_0)^{\frac{1}{m-1}}} ds < +\infty.$$

Choosing then various functions ϕ, which, together with φ and u, satisfy the conditions (c1), (c2) and (c4) of Corollary 5.7, one gets more concrete results. Possible choices are e.g.

$$\phi(t) = \frac{t^p}{(\log t)^q},\ p > 0, q \in \mathbb{R}.$$

The case $p = 1$ and $q = 0$ gives the result of Armitage and Gardiner.

Example 12. Let u be separately subharmonic in Ω and $\varphi : [0,+\infty) \to [0,+\infty)$ be a strictly increasing surjection. Let $p > 0$, $q \geq 0$. Let $\phi : [0,+\infty) \to [0,+\infty)$ be a strictly increasing surjection for which there is $t_1 > 1$ such that

$$s = \phi(t) = e^{\left(\frac{\log t}{p}\right)^{\frac{1}{q+1}}} = e^{\sqrt[q+1]{\frac{\log t}{p}}} \quad \text{for } t \geq t_1,$$

thus $t = \phi^{-1}(s) = e^{p(\log s)^{q+1}}$. One sees easily that the condition (c1) of Corollary 5.7 is satisfied, but the condition (c2) not. As a matter of fact, and as one easily sees,

$$\frac{\phi^{-1}(e^{s+2})}{\phi^{-1}(e^s)} \to +\infty \text{ as } s \to +\infty.$$

Therefore, in this case one cannot use Corollary 5.7 to conclude that u is subharmonic. However, using Corollary 5.6 we see that u is subharmonic, provided that

$$(5.2) \qquad \int_{s_5}^{+\infty} \frac{e^{\frac{p}{m-1}[(s+2)^{q+1}-s^{q+1}]}}{\varphi(s-s_0)^{\frac{1}{m-1}}} \left(\int_{s_5+s_0+2}^{s+s_0+2} e^{\frac{p}{n-1}[(t+2)^{q+1}-t^{q+1}]} dt \right)^{\frac{n-1}{m-1}} ds < +\infty,$$

and (this is just (c3))

$$\varphi \circ \left(\left[\frac{\log^+ u}{p} \right]^{\frac{1}{q+1}} \right) \in \mathcal{L}^1_{\text{loc}}(\Omega).$$

The condition (5.2) is of course complicated, but it is easy to get simpler (but) stronger conditions, e.g. just estimating the inner integral.

Example 13. Let u be separately subharmonic in Ω and $\varphi : [0,+\infty) \to [0,+\infty)$ be a strictly increasing surjection. Let $p > 0$ and $\phi : [0,+\infty) \to [0,+\infty)$ be a strictly increasing surjection for which there is $t_1 > 1$ such that

$$s = \phi(t) = (\log t)^p \quad \text{for } t \geq t_1,$$

and thus

$$t = \phi^{-1}(s) = e^{s^{\frac{1}{p}}}.$$

Corollary 5.7 cannot now be applied, but from Corollary 5.6 it follows that u is subharmonic, provided that, in addition to the integrability condition (c3),

$$\varphi \circ \log^+ \circ ((\log^+ u^+)^p) \in \mathcal{L}^1_{\text{loc}}(\Omega),$$

also the condition (c2) holds. One possibility to replace the, again rather complicated, condition (c2) by a simpler, but stronger one, is the following (we leave the details to the reader):

$$\int_{s_5}^{+\infty} e^{\frac{2}{m-1}[e^{\frac{1}{p}(s+s_0+4)} - e^{\frac{1}{p}(s+s_0+2)}]} \frac{s^{\frac{n-1}{m-1}}}{\varphi(s-s_0)^{\frac{1}{m-1}}} ds < +\infty.$$

5.3. Improvements to the basic concise results of Lelong, Avanissian, Arsove and ours.

5.3.1. ***Motivation***. As was seen above, the proof of Theorem 5.5 is based on a previous simple result of separately quasinearly subharmonic functions, namely on [109], Proposition 3.1, p. 57, see also [113], Proposition 3.2.1, p. e2620, and Proposition 5.4 below. The situation is of course similar in the special case of separately subharmonic functions: Armitage and Gardiner [3], proof of Theorem 1, pp. 257-259, base their result on the classical result of Avanissian, [6], Théorème 9, p. 140 (the proof of which, however, is not at all simple; see also the first result on the area, namely Lelong, [65], Théorème 1 bis, p. 315). Equally well one might of course base the result on any of the following later results: [67], Proposition 3, p. 24, [49], Theorem, p. 31, Arsove [5], Theorem 1, p. 622, [94], Theorem 1, p. 69, Cegrell and Sadullaev [14], Theorem 3.1, p. 82.

Since the result of Armitage and Gardiner and especially our improvement, Theorem 5.5 above, are both somewhat complicated, it is worthwhile to give improvements also to the above cited basic and concise results of Lelong and Avanissian, of Arsove and of ours. The methods and ideas of the below presented proofs have their roots already in [94], see also [106, 107]. The proofs are rather simple, especially when compared with the proofs of the already cited old results.

5.3.2. ***A concise result for separately quasinearly subharmonic functions***. In the proof of our result below, we need the following simple result:

Proposition 5.8. ([109], Proposition 3.1, p. 57, and [113], Proposition 3.2.1, p. e2620) *Let Ω be a domain in $\mathbb{R}^{m+n}$, $m,n\geq 2$, and let $K_1,K_2\geq 1$. Let $u:\Omega\to[-\infty,+\infty)$ be a Lebesgue measurable function such that*

(a) *for each $y\in\mathbb{R}^n$ the function*

$$\Omega(y)\ni x\mapsto u(x,y)\in[-\infty,+\infty)$$

is K_1-quasinearly subharmonic,

(b) *for almost every $x\in\mathbb{R}^m$ the function*

$$\Omega(x)\ni y\mapsto u(x,y)\in[-\infty,+\infty)$$

is K_2-quasinearly subharmonic,

(c) *there exists a non-constant permissible function $\psi:[0,+\infty)\to[0,+\infty)$ such that $\psi\circ u^+\in\mathcal{L}^1_{\rm loc}(\Omega)$.*

Then u is $\frac{4^{m+n}\nu_{m+n}K_1K_2}{\nu_m\nu_n}$-quasinearly subharmonic in Ω.

Theorem 5.9. ([116], Theorem 1, pp. 365–366) *Let Ω be a domain in $\mathbb{R}^{m+n}$, $m,n\geq 2$, and let $K\geq 1$. Let $u:\Omega\to[-\infty,+\infty)$ be a Lebesgue measurable function such that*

(a) *for each $y\in\mathbb{R}^n$ the function*

$$\Omega(y)\ni x\mapsto u(x,y)\in[-\infty,+\infty)$$

is 1-quasinearly subharmonic,

(b) *for almost every* $x \in \mathbb{R}^m$ *the function*

$$\Omega(x) \ni y \mapsto u(x,y) \in [-\infty, +\infty)$$

is K-quasinearly subharmonic,

(c) *there exists a non-constant permissible function* $\psi : [0,+\infty) \to [0,+\infty)$ *such that* $\psi \circ u^+ \in \mathcal{L}^1_{\mathrm{loc}}(\Omega)$.

Then u is K-quasinearly subharmonic in Ω.

Proof. By Proposition 5.8 u is quasinearly subharmonic and thus locally integrable in Ω.

Take $M \geq 0$ arbitrarily, and write $u_M := \max\{u, -M\} + M$. It remains to show that for all $(a,b) \in \Omega$ and $R > 0$ such that $\overline{B^{m+n}((a,b),R)} \subset \Omega$,

$$u_M(a,b) \leq \frac{K}{\nu_{m+n}R^{m+n}} \int\limits_{B^{m+n}((a,b),R)} u_M(x,y)\,dm_{m+n}(x,y).$$

To see this, we just proceed in the following standard way, see e.g. [49], proof of Theorem a), pp. 32-33:

$$\frac{K}{\nu_{m+n}R^{m+n}} \int\limits_{B^{m+n}((a,b),R)} u_M(x,y)\,dm_{m+n}(x,y) =$$

$$= \frac{\nu_n}{\nu_{m+n}R^{m+n}} \int\limits_{B^m(a,R)} [(R^2 - |x-a|^2)^{\frac{n}{2}} \times$$

$$\times \frac{K}{\nu_n(R^2 - |x-a|^2)^{\frac{n}{2}}} \int\limits_{B^n(b,\sqrt{R^2-|x-a|^2})} u_M(x,y)\,dm_n(y)]\,dm_m(x) \geq$$

$$\geq \frac{\nu_n}{\nu_{m+n}R^{m+n}} \int\limits_{B^m(a,R)} (R^2 - |x-a|^2)^{\frac{n}{2}} u_M(x,b)\,dm_m(x) \geq u_M(a,b).$$

Above we have used, in addition to the fact that, for almost every $x \in \mathbb{R}^m$, the functions $u_M(x,\cdot)$ are K-quasinearly subharmonic, also the following lemma. (Observe that the proof of the Lemma, see [49], proof of Theorem 2 a), p. 15, works also in our slightly more general situation.)

Lemma 5.10. ([49], Theorem 2 a), p. 15) *Let v be nearly subharmonic (in the generalized sense, defined above) in a domain U of* $\mathbb{R}^N$, $N \geq 2$, $\psi \in \mathcal{L}^\infty(\mathbb{R}^N)$, $\psi \geq 0$, $\psi(x) = 0$ *when* $|x| \geq \alpha$ *and* $\psi(x)$ *depends only on* $|x|$. *Then* $\psi \star v \geq v$ *and* $\psi \star v$ *is subharmonic in* U_α, *provided* $\int \psi(x)\,dm_N(x) = 1$, *where* $U_\alpha = \{x \in U : \overline{B^N(x,\alpha)} \subset U\}$.

□

Remark 5.11. The above result improves our previous result [109], Theorem 3.1, p. 58, and [113], Theorem 3.2.2, p. e2620, by replacing, among others, the previously used quasinearly subharmonic n.s. (quasinearly subharmonic in the narrow sense) functions with quasinearly subharmonic functions.

Corollary 5.12. ([109], Corollary 3.1, p. 59, and [113], Corollary 3.2.3, pp. e2620–e2621) *Let Ω be a domain in $\mathbb{R}^{m+n}$, $m,n \geq 2$. Let $u : \Omega \to [-\infty,+\infty)$ be a Lebesgue measurable function such that*

(a) *for each $y \in \mathbb{R}^n$ the function*

$$\Omega(y) \ni x \mapsto u(x,y) \in [-\infty,+\infty)$$

is nearly subharmonic,

(b) *for almost every $x \in \mathbb{R}^m$ the function*

$$\Omega(x) \ni y \mapsto u(x,y) \in [-\infty,+\infty)$$

is nearly subharmonic,

(c) *there exists a non-constant permissible function $\psi : [0,+\infty) \to [0,+\infty)$ such that $\psi \circ u^+ \in \mathcal{L}_{\mathrm{loc}}(\Omega)$.*

Then u is nearly subharmonic in Ω.

5.3.3. ***A concise result for separately subharmonic functions***. We begin with an improvement to [109], Corollary 3.2, p. 61, and [113], Corollary 3.2.4, p. e2621:

Theorem 5.13. *Let Ω be a domain in $\mathbb{R}^{m+n}$, $m,n \geq 2$, and let $K_1, K_2 \geq 1$. Let $u : \Omega \to [-\infty,+\infty)$ be such that*

(a) *for each $y \in \mathbb{R}^n$ the function*

$$\Omega(y) \ni x \mapsto u(x,y) \in [-\infty,+\infty)$$

is K_1-quasinearly subharmonic, and, for almost every $y \in \mathbb{R}^n$, subharmonic,

(b) *for each $x \in \mathbb{R}^m$ the function*

$$\Omega(x) \ni y \mapsto u(x,y) \in [-\infty,+\infty)$$

is upper semicontinuous, and, for almost every $x \in \mathbb{R}^m$, K_2-quasinearly subharmonic,

(c) *there exists a non-constant permissible function $\psi : [0,+\infty) \to [0,+\infty)$ such that $\psi \circ u^+ \in \mathcal{L}^1_{\mathrm{loc}}(\Omega)$.*

Then for each $(a,b) \in \Omega$,

$$\limsup_{(x,y)\to(a,b)} u(x,y) \leq K_1 K_2 u^+(a,b).$$

Proof. By [109], Lemma, pp. 59–60, u is measurable. By Proposition 5.8, u and thus also u^+ are quasinearly subharmonic and thus locally bounded above. Clearly u^+ satisfies the assumptions of the theorem. It is sufficient to show that for any $(a,b) \in \Omega$,

$$\limsup_{(x,y)\to(a,b)} u^+(x,y) \leq K_1 K_2 u^+(a,b).$$

Take $(a,b)\in\Omega$ and $R_1>0$ and $R_2>0$ arbitrarily such that $\overline{B^m(a,R_1)\times B^n(b,R_2)}\subset\Omega$. Choose an arbitrary $\lambda\in\mathbb{R}$ such that $u^+(a,b)<\lambda$. Since $u^+(a,\cdot)$ is upper semicontinuous, we find R_2', $0<R_2'<R_2$, such that

$$\frac{1}{\nu_n R_2'^n}\int\limits_{B^n(b,R_2')} u^+(a,y)dm_n(y)<\lambda.$$

Using the fact that, for almost every $y\in\mathbb{R}^n$, the function $u^+(\cdot,y)$, is subharmonic, we get

$$\frac{1}{\nu_m r^m}\int\limits_{B^m(a,r)} u^+(x,y)dm_m(x)\to u^+(a,y)\ \text{ as } r\to 0.$$

Since u^+ is locally bounded above, one can use the Lebesgue Dominated Convergence Theorem. Thus we find R_1', $0<R_1'<R_1$, such that

$$\frac{1}{\nu_n R_2'^n}\int\limits_{B^n(b,R_2')}\Big[\frac{1}{\nu_m R_1'^m}\int\limits_{B^m(a,R_1')} u^+(x,y)dm_m(x)\Big]dm_n(y)<\lambda.$$

Choose r_1, $0<r_1<R_1'$, and r_2, $0<r_2<R_2'$, arbitrarily. Then for each $(x,y)\in B^m(a,r_1)\times B^n(b,r_2)$,

$$u^+(x,y)\le\frac{K_1}{\nu_m(R_1'-r_1)^m}\int\limits_{B^m(x,R_1'-r_1)} u^+(\xi,y)dm_m(\xi)\le$$

$$\le\frac{K_1}{\nu_m(R_1'-r_1)^m}\int\limits_{B^m(x,R_1'-r_1)}\Big[\frac{K_2}{\nu_n(R_2'-r_2)^n}\int\limits_{B^n(y,R_2'-r_2)} u^+(\xi,\eta)dm_n(\eta)\Big]dm_m(\xi)\le$$

$$\le\frac{K_2}{\nu_n(R_2'-r_2)^n}\int\limits_{B^n(y,R_2'-r_2)}\Big[\frac{K_1}{\nu_m(R_1'-r_1)^m}\int\limits_{B^m(x,R_1'-r_1)} u^+(\xi,\eta)dm_m(\xi)\Big]dm_n(\eta)\le$$

$$\le\left(\frac{R_1'}{R_1'-r_1}\right)^m\cdot\left(\frac{R_2'}{R_2'-r_2}\right)^n\cdot K_1K_2\frac{1}{\nu_n R_2'^n}\int\limits_{B^n(b,R_2')}\Big[\frac{1}{\nu_m R_1'^m}\int\limits_{B^m(a,R_1')} u^+(\xi,\eta)dm_m(\xi)\Big]dm_n(\eta)<$$

$$<\left(\frac{R_1'}{R_1'-r_1}\right)^m\cdot\left(\frac{R_2'}{R_2'-r_2}\right)^n\cdot K_1K_2\cdot\lambda.$$

Sending then $r_1\to 0$, $r_2\to 0$, one gets

$$\limsup_{(x,y)\to(a,b)} u^+(x,y)\le K_1K_2\cdot\lambda,$$

concluding the proof. □

Observe that the proof of the above (quasinearly subharmonicity) result, and thus also the proof of the following special case result, is simpler than the proofs of the older subharmonicity results.

Corollary 5.14. ([117], Corollary 2, p. 364) *Let Ω be a domain in $\mathbb{R}^{m+n}$, $m,n \geq 2$. Let $u : \Omega \to [-\infty,+\infty)$ be such that*

(a) *for each $y \in \mathbb{R}^n$ the function*

$$\Omega(y) \ni x \mapsto u(x,y) \in [-\infty,+\infty)$$

is nearly subharmonic, and, for almost every $y \in \mathbb{R}^n$, subharmonic,

(b) *for each $x \in \mathbb{R}^m$ the function*

$$\Omega(x) \ni y \mapsto u(x,y) \in [-\infty,+\infty)$$

is upper semicontinuous, and, for almost every $x \in \mathbb{R}^m$, (nearly) subharmonic,

(c) *for some $p > 0$ there is a function $v \in \mathcal{L}^p_{\mathrm{loc}}(\Omega)$ such that $u \leq v$.*

Then u is upper semicontinuous and thus subharmonic in Ω.

Proof. It is easy to see that for each $M \geq 0$, the function $u_M := \max\{u,-M\}+M$ satisfies the assumptions of Theorem 5.9. Thus u_M is upper semicontinuous. Since by Corollary 5.12, u_M is anyway nearly subharmonic, it is in fact subharmonic. Using then e.g. [49], a), p. 8, one sees that u is subharmonic and thus also upper semicontinuous. □

Remark 5.15. Observe that Corollary 5.14 is partially related to the result [49], Proposition 2, pp. 34–35: Though our assumptions are slightly stronger, our proof is, on the other hand, different and shorter.

Corollary 5.16. ([94], Theorem 1, p. 69) *Let Ω be a domain in $\mathbb{R}^{m+n}$, $m,n \geq 2$. Let $u : \Omega \to [-\infty,+\infty)$ be such that*

(a) *for each $y \in \mathbb{R}^n$ the function*

$$\Omega(y) \ni x \mapsto u(x,y) \in [-\infty,+\infty)$$

is subharmonic,

(b) *for each $x \in \mathbb{R}^m$ the function*

$$\Omega(x) \ni y \mapsto u(x,y) \in [-\infty,+\infty)$$

is subharmonic,

(c) *for some $p > 0$ there is a function $v \in \mathcal{L}^p_{\mathrm{loc}}(\Omega)$ such that $u \leq v$.*

Then u is subharmonic in Ω.

6. Separately Subharmonic and Harmonic Functions

Abstract. It is an open problem whether a function, subharmonic with respect to the first variable and harmonic with respect to the second, is subharmonic or not. Based again on our mean value type inequality, we improve our previous subharmonicity results of the above type functions, thus improving also the previous results of Kołodziej and Thorbiörnson and Imomkulov. Moreover, we give refinements, with concise proofs, to the older basic results of Arsove, and of Cegrell and Sadullaev.

Keywords. Subharmonic, quasinearly subharmonic, separately quasinearly subharmonic and harmonic

6.1. **Previous results.** As pointed out already above, Wiegerinck [155], Theorem, p. 770, see also Wiegerinck and Zeinstra [156], Theorem 1, p. 246, showed that a separately subharmonic function need not be subharmonic. On the other hand, it is an open problem, whether a function which is subharmonic in one variable and harmonic in the other, is subharmonic. For older basic results on this area, see e.g. Arsove [5], Theorem 2, p. 622, Imomkulov [52], Theorem, p. 9, Wiegerinck and Zeinstra [156], p. 248, Cegrell and Sadullaev [14], Theorem 3.1, p. 82, and Kołodziej and Thorbiörnson [61], Theorem 1, p. 463. See Sadullaev [130] and Kalantar [56] for related recent results on the area.

The result of Kołodziej and Thorbiörnson includes the results of Arsove, of Cegrell and Sadullaev and of Imomkulov, and reads as follows:

Theorem 6.1. *Let Ω be a domain in $\mathbb{R}^{m+n}$, $m,n \geq 2$. Let $u : \Omega \to \mathbb{R}$ be such that*

(a) *for each $y \in \mathbb{R}^n$ the function*

$$\Omega(y) \ni x \mapsto u(x,y) \in \mathbb{R}$$

is subharmonic and $\mathcal{C}^2$,

(b) *for each $x \in \mathbb{R}^m$ the function*

$$\Omega(x) \ni y \mapsto u(x,y) \in \mathbb{R}$$

is harmonic.

Then u is subharmonic and continuous in Ω.

We improved the result of Kołodziej and Thorbiörnson in a series of papers: [106], Theorem 3, Theorem 4 and Corollary, pp. 162–164, [107], Theorem 6, p. 234, [108], Theorem 1 and Corollary, pp. 438, 444, [109], Theorem 5.1, Theorem 5.2, Corollary 5.1 and Corollary 5.2, pp. 67, 72–74, see also [113], Theorem 4.3.1, Theorem 4.3.2, Corollary 4.3.3 and Corollary 4.3.4, pp. e2625–e2626. We will now return to the subject and improve our result still further, see Theorem 6.5 below.

However, we begin with improving the above cited results of Arsove and of Cegrell and Sadullaev and our previous generalizations. Instead of subharmonic functions (resp. so called quasinearly subharmonic functions n.s.), we will now use quasinearly subharmonic functions. Again this is indeed useful.

Juhani Riihentaus

6.2. **Arsove's result.** Arsove's result is:

Theorem 6.2. ([5], Theorem 2, p. 622) *Let Ω be a domain in $\mathbb{R}^{m+n}$, $m,n \geq 2$. Let $u : \Omega \to \mathbb{R}$ be such that*

(a) *for each $y \in \mathbb{R}^n$ the function*

$$\Omega(y) \ni x \mapsto u(x,y) \in \mathbb{R}$$

is subharmonic,

(b) *for each $x \in \mathbb{R}^m$ the function*

$$\Omega(x) \ni y \mapsto u(x,y) \in \mathbb{R}$$

is harmonic,

(c) *there is a nonnegative function $\varphi \in \mathcal{L}^1_{\mathrm{loc}}(\Omega)$ such that $-\varphi \leq u$.*

Then u is subharmonic in Ω.

Arsove's proof was based on mean value operators. Much later Cegrell and Sadullaev [14], Theorem 3.1, p. 82, gave a new proof using Poisson modification.

6.3. **Improvements to Arsove's result and to Cegrell's and Sadullaev's result.** Below in Theorem 6.4 we generalize the above result of Arsove, and of Cegrell and Sadullaev, and also our previous improvements [109], Theorem 4.1, Theorem 4.2 and Corollary 4.1, pp. 64–65, see also [113], Theorem 4.2.1, Theorem 4.2.3 and Corollary 4.2.2, p. e2623–e2624. Our proof is direct and short, and it is different than the previous proofs of Arsove and of Cegrell and Sadullaev. In the proof we need the following lemma.

Lemma 6.3. *Let D be a domain in $\mathbb{R}^N$, $N \geq 2$. Let E be a locally compact space, and let μ be a positive measure on E. Let $v : D \times E \to \mathbb{R}$ be such that*

(a) *for each $y \in E$ the function*

$$D \ni x \mapsto v(x,y) \in \mathbb{R}$$

is continuous,

(b) *for each $x \in D$ the function*

$$E \ni y \mapsto v(x,y) \in \mathbb{R}$$

is measurable.

Then v is measurable.

The idea of the proof is just to use dyadic cubes and approximate the function $v(\cdot,\cdot)$ with measurable functions of the type

$$D \times E \ni (x,y) \mapsto v(a,y) \in \mathbb{R}, \quad a \in D.$$

We leave the details to the reader and just refer to a preliminary result [127], Exercise 8, p. 160, and to the special case results [5], Lemma 1, p. 624, and [92], Lemma 3.2, pp. 103–104, say.

Theorem 6.4. *Let Ω be a domain in $\mathbb{R}^{m+n}$, $m,n \geq 2$, and $K \geq 1$. Let $u : \Omega \to \mathbb{R}$ be such that*

(a) *for each $y \in \mathbb{R}^n$ the function*

$$\Omega(y) \ni x \mapsto u(x,y) \in \mathbb{R}$$

is K-quasinearly subharmonic,

(b) *for each $x \in \mathbb{R}^m$ the function*

$$\Omega(x) \ni y \mapsto u(x,y) \in \mathbb{R}$$

is harmonic,

(c) *there is a nonnegative function $\varphi \in \mathcal{L}^1_{\mathrm{loc}}(\Omega)$ such that $-\varphi \leq u$.*

Then u is K-quasinearly subharmonic in Ω.

Proof. The Lebesgue measurability of u follows from Lemma 6.3. Thus also the functions $u_M := \max\{u, -M\} + M$, $M \geq 0$, are Lebesgue measurable. We must show that $u^+ \in \mathcal{L}^1_{\mathrm{loc}}(\Omega)$ and that each u_M satisfies the generalized mean value inequality.

To see that $u^+ \in \mathcal{L}^1_{\mathrm{loc}}(\Omega)$, we proceed as follows. Observe first that $0 \leq u^+ \leq u_M \leq v_M := u + \varphi + M$. To see that $v_M \in \mathcal{L}^1_{\mathrm{loc}}(\Omega)$ requires *only* Fubini's theorem. As a matter of fact, take $\overline{B^m(a,R) \times B^n(b,R)} \subset \Omega$ arbitrarily.

Then

$$0 \leq \frac{K}{m_{m+n}(B^m(a,R) \times B^n(b,R))} \int\limits_{B^m(a,R)\times B^n(b,R)} v_M(x,y)\,dm_{m+n}(x,y) \leq$$

$$\leq \frac{K}{m_{m+n}(B^m(a,R) \times B^n(b,R))} \int\limits_{B^m(a,R)\times B^n(b,R)} [u(x,y) + \varphi(x,y) + M]\,dm_{m+n}(x,y) \leq$$

$$\leq \frac{K}{\nu_m R^m} \int\limits_{B^m(a,R)} \{\frac{1}{\nu_n R^n} \int\limits_{B^n(b,R)} [u(x,y) + \varphi(x,y) + M]\,dm_n(y)\}\,dm_m(x) \leq$$

$$\leq \frac{K}{\nu_m R^m} \int\limits_{B^m(a,R)} [\frac{1}{\nu_n R^n} \int\limits_{B^n(b,R)} u(x,y)\,dm_n(y) + \frac{1}{\nu_n R^n} \int\limits_{B^n(b,R)} \varphi(x,y)\,dm_n(y) + M]\,dm_m(x) \leq$$

$$\leq \frac{K}{\nu_m R^m} \int\limits_{B^m(a,R)} [u(x,b) + \frac{1}{\nu_n R^n} \int\limits_{B^n(b,R)} \varphi(x,y)\,dm_n(y) + M]\,dm_m(x) \leq$$

$$\leq \frac{K}{\nu_m R^m} \int_{B^m(a,R)} u(x,b)dm_m(x) +$$

$$+\frac{K}{\nu_m R^m} \int_{B^m(a,R)} [\frac{1}{\nu_n R^n} \int_{B^n(b,R)} \varphi(x,y)dm_n(y)]dm_m(x) + KM \leq$$

$$\leq \frac{K}{\nu_m R^m} \int_{B^m(a,R)} u(x,b)dm_m(x) +$$

$$+ \frac{K}{m_{m+n}(B^m(a,R) \times B^n(b,R))} \int_{B^m(a,R)\times B^n(b,R)} \varphi(x,y)dm_{m+n}(x,y) + KM <$$

$$< +\infty.$$

It remains to show that for all $(a,b) \in \Omega$ and $R > 0$ such that $\overline{B^{m+n}((a,b),R)} \subset \Omega$,

$$u_M(a,b) \leq \frac{K}{\nu_{m+n}R^{m+n}} \int_{B^{m+n}((a,b),R)} u_M(x,y)dm_{m+n}(x,y).$$

To see this, we proceed in the following standard, direct and short way, see e.g. [49], Proposition 2 c) and proof of Theorem a), pp. 10–11, 32–33, and [109], p. 59:

$$\frac{K}{\nu_{m+n}R^{m+n}} \int_{B^{m+n}((a,b),R)} u_M(x,y)dm_{m+n}(x,y) =$$

$$= \frac{\nu_m}{\nu_{m+n}R^{m+n}} \int_{B^n(b,R)} [(R^2 - \mid y-b \mid^2)^{\frac{m}{2}} \frac{K}{\nu_m (R^2 - \mid y-b \mid^2)^{\frac{m}{2}}} \times$$

$$\times \int_{B^m(a,\sqrt{R^2 - |y-b|^2})} u_M(x,y)dm_m(x)]dm_n(y) \geq$$

$$\geq \frac{\nu_m}{\nu_{m+n}R^{m+n}} \int_{B^n(b,R)} (R^2 - \mid y-b \mid^2)^{\frac{m}{2}} u_M(a,y)dm_n(y) \geq u_M(a,b).$$

Above we have used, in addition to the fact that, for every $y \in \mathbb{R}^m$, the functions $u(\cdot,y)$ are K-quasinearly subharmonic, also the above Lemma 5.10. □

6.4. **An improvement to the result of Kołodziej and Thorbiörnson.** In our generalization to the result of Kołodziej and Thorbiörnson, we will use the generalized Laplacian, defined with the aid of the Blaschke-Privalov operators, see e.g. [131], p. 451, [126], pp. 278–279, [132], p. 498, and [134], p. 29. Let D be a domain in $\mathbb{R}^N$,

$N \geq 2$, and $f : D \to \mathbb{R}$, $f \in \mathcal{L}^1_{\mathrm{loc}}(D)$. We write

$$\Delta_* f(x) := \liminf_{r \to 0} \frac{2(N+2)}{r^2} \cdot \Big[\frac{1}{\nu_N r^N} \int_{B^N(x,r)} f(x')dm_N(x') - f(x)\Big],$$

$$\Delta^* f(x) := \limsup_{r \to 0} \frac{2(N+2)}{r^2} \cdot \Big[\frac{1}{\nu_N r^N} \int_{B^N(x,r)} f(x')dm_N(x') - f(x)\Big].$$

If $\Delta_* f(x) = \Delta^* f(x)$, then write $\Delta f(x) := \Delta_* f(x) = \Delta^* f(x)$.

If $f \in \mathcal{C}^2(D)$, then

$$\Delta f(x) = \Big(\sum_{j=1}^{N} \frac{\partial^2 f}{\partial x_j^2}\Big)(x),$$

the standard Laplacian with respect to the variable $x = (x_1, x_2, \ldots, x_N)$. More generally, if $x \in D$ and $f \in t_2^1(x)$, i.e. f has an $\mathcal{L}^1$ total differential of order 2 at x, then $\Delta f(x)$ equals with the pointwise Laplacian of f at x, i.e.

$$\Delta f(x) = \sum_{j=1}^{N} D_{jj} f(x).$$

Here $D_{jj} f$ represents a generalization to the usual $\frac{\partial^2 f}{\partial x_j^2}$, $j = 1, 2, \ldots, N$. See e.g. [133], p. 369, and [134], p. 29.

Recall that there exist functions which are not $\mathcal{C}^2$ but for which the generalized Laplacian is nevertheless continuous, perhaps in the extended sense (in $([0, +\infty], q)$, where q is the spherical metric). See the examples below (and [134], p. 31, [109], Example 5 and Example 6, pp. 66–67, and [113], Example 1 and Example 2, pp. e2624–e2625).

Example 14. ([134], p. 31) The function $f : \mathbb{R}^N \to \mathbb{R}$,

$$f(x) := \begin{cases} -1, & \text{when } x_N < 0, \\ 0, & \text{when } x_N = 0, \\ 1, & \text{when } x_N > 0, \end{cases}$$

is non-continuous, but nevertheless $\Delta f(x) = 0$ for all $x \in \mathbb{R}^N$.

Example 15. Let $1 \leq k \leq N$ and let $E = \{(0, 0, \ldots, 0)\} \times \mathbb{R}^{N-k} \subset \mathbb{R}^N$. Let $0 < \lambda \leq 1$. Define $f : \mathbb{R}^N \to \mathbb{R}$,

$$f(x) = f(x_1, x_2, \ldots, x_k, x_{k+1}, \ldots, x_N) := \left(\sqrt{x_1^2 + x_2^2 + \cdots + x_k^2}\right)^{\lambda}.$$

Then f is continuous and subharmonic in $\mathbb{R}^N$, but not in $\mathcal{C}^1(\mathbb{R}^N)$. Nevertheless, Δf is defined everywhere in $\mathbb{R}^N$, equals $+\infty$ in E, and is continuous in $\mathbb{R}^N$, in E in the extended sense, with respect to the spherical metric:

$$q(a,b) := \begin{cases} \frac{|a-b|}{\sqrt{1+a^2}\sqrt{1+b^2}} & \text{when } a,b \in [0,+\infty), \\ \frac{1}{\sqrt{1+a^2}} & \text{when } a \in [0,+\infty) \text{ and } b = +\infty. \end{cases}$$

Observe that $([0,+\infty],q)$ is a complete metric space.

If f is subharmonic on D, it follows from [131], p. 451, (see also [126], Lemma 2.2, p. 280, and [133], p. 376) that $\Delta f(x) = \Delta_* f(x) = \Delta^* f(x) \in \mathbb{R}$ for almost every $x \in D$.

Below we use the following notation. Let Ω is a domain in $\mathbb{R}^{m+n}$, $m,n \geq 2$, and $u : \Omega \to \mathbb{R}$. If $y \in \mathbb{R}^n$ is such that the function

$$\Omega(y) \ni x \mapsto f(x) := u(x,y) \in \mathbb{R}$$

is in $\mathcal{L}^1_{\mathrm{loc}}(\Omega(y))$, then we write $\Delta_{1*}u(x,y) := \Delta_* f(x)$, $\Delta_1^* u(x,y) := \Delta^* f(x)$, and $\Delta_1 u(x,y)$ $:= \Delta f(x)$.

Then our generalization to our previous result [109], Theorem 5.1, pp. 67–68, (or [113], Theorem 4.3.1, p. e2625 (where no proof was given!)) and thus also the result of Kołodziej and Thorbiörnson [61], Theorem 1, p. 463, Theorem 6.1 above. Though our proof will follow the main lines of [109], proof of Theorem 5.1, pp. 67–72, it is different enough, nevertheless, to warrant that it be given in complete detail here. As a matter of fact, our assumption (d) is now essentially milder than our previous assumptions (d) and (e).

Theorem 6.5. *Let Ω be a domain in $\mathbb{R}^{m+n}$, $m,n \geq 2$. Let $u : \Omega \to \mathbb{R}$ be such that for each $(x',y') \in \Omega$ there is $(x_0,y_0) \in \Omega$ and $r_1 > 0$, $r_2 > 0$ such that $(x',y') \in B^m(x_0,r_1) \times B^n(y_0,r_2) \subset \overline{B^m(x_0,r_1) \times B^n(y_0,r_2)} \subset \Omega$ and such that the following conditions are satisfied:*

(a) *For each $y \in \overline{B^n(y_0,r_2)}$ the function*

$$\overline{B^m(x_0,r_1)} \ni x \mapsto u(x,y) \in \mathbb{R}$$

is continuous, and subharmonic in $B^m(x_0,r_1)$.

(b) *For each $x \in \overline{B^m(x_0,r_1)}$ the function*

$$\overline{B^n(y_0,r_2)} \ni y \mapsto u(x,y) \in \mathbb{R}$$

is continuous, and harmonic in $B^n(y_0,r_2)$.

(c) *For each $y \in B^n(y_0,r_2)$ one has $\Delta_{1*}u(x,y) < +\infty$ for each $x \in B^m(x_0,r_1)$, possibly with the exception of a polar set in $B^m(x_0,r_1)$.*

(d) *There are a set $H_1 \subset B^m(x_0,r_1)$, dense in $B^m(x_0,r_1)$, and a set $H_2 \subset B^n(y_0,r_2)$, dense in $B^n(y_0,r_2)$, such that*

(d1) *for each $y \in H_2$, for almost every $x \in B^m(x_0,r_1)$ and for each $x \in H_1$,*

$$\Delta_1 u(x',y) \to \Delta_1 u(x,y) \in \mathbb{R}$$

as $x' \to x$, $x' \in H_1$, and

(d2) *for each* $y \in B^n(y_0,r_2) \setminus H_2$ *and for almost every* $x \in B^m(x_0,r_1)$,

$$\Delta_1 u(x,y') \to \Delta_1 u(x,y) \in \mathbb{R}$$

as $y' \to y$, $y' \in H_2$.

Then u is subharmonic in Ω.

Proof. Choose r_1', r_2' such that $0 < r_1' < r_1$, $0 < r_2' < r_2$, and such that $(x',y') \in B^m(x_0,r_1') \times B^n(y_0,r_2')$. It is sufficient to show that $u \mid B^m(x_0,r_1') \times B^n(y_0,r_2')$ is subharmonic. For the sake of convenience of notation, we change the roles of r_j and r_j', $j = 1,2$. We divide the proof into several steps.

Step 1 *Construction of an auxiliar dense set G.*

For each $k \in \mathbb{N}$ write

$$A_k := \{x \in \overline{B^m(x_0,r_1)} : -k \le u(x,y) \le k \quad \text{for each} \quad y \in \overline{B^n(y_0,r_2)}\}.$$

Clearly A_k is closed, and

$$\overline{B^m(x_0,r_1)} = \bigcup_{k=1}^{+\infty} A_k.$$

Write

$$G := \bigcup_{k=1}^{+\infty} \operatorname{int} A_k.$$

It follows from Baire's theorem that G is dense in $B^m(x_0,r_1)$.

Step 2 *The functions* $\Delta_{1r}u(x,\cdot) : B^n(y_0,r_2) \to \mathbb{R}$ *(see the definition below),* $x \in G$, $0 < r < r_x := \operatorname{dist}(x, \overline{B^m(x_0,r_1)} \setminus G)$, *are nonnegative and harmonic.*

For each $(x,y) \in B^m(x_0,r_1) \times B^n(y_0,r_2)$ and each r, $0 < r < \operatorname{dist}(x, \partial B^m(x_0,r_1'))$ (observe that $\operatorname{dist}(x, \partial B^m(x_0,r_1')) > r_1' - r_1 > 0$), write

$$\Delta_{1r}u(x,y) := \frac{2(m+2)}{r^2} \cdot \Big[\frac{1}{\nu_m r^m} \int_{B^m(x,r)} u(x',y)\, dm_m(x') - u(x,y)\Big] =$$

$$= \frac{2(m+2)}{r^2} \cdot \frac{1}{\nu_m r^m} \int_{B^m(0,r)} \big[u(x+x',y) - u(x,y)\big]\, dm_m(x').$$

Since $u(\cdot,y)$ is subharmonic, $\Delta_{1r}u(x,y)$ is defined and nonnegative. Suppose then that $x \in G$ and $0 < r < r_x$. Since $\overline{B^m(x,r)} \subset G$ and $A_k \subset A_{k+1}$ for all $k = 1,2,\ldots$, $\overline{B^m(x,r)} \subset \operatorname{int} A_N$ for some $N \in \mathbb{N}$. Therefore

$$-N \le u(x',y) \le N \text{ for all } x' \in B^m(x,r) \text{ and } y \in B^n(y_0,r_2),$$

and hence

$$\text{(6.1)} \qquad -2N \le u(x+x',y) - u(x,y) \le 2N \text{ for all } x' \in B^m(0,r) \text{ and } y \in B^n(y_0,r_2).$$

To show that $\Delta_{1r}u(x,\cdot)$ is continuous, pick an arbitrary sequence $y_j \to \tilde{y}_0$, $y_j, \tilde{y}_0 \in B^n(y_0,r_2)$, $j = 1,2,\ldots$. Using then (6.1), the Lebesgue Dominated Convergence Theorem and the continuity of $u(x,\cdot)$, one sees easily that $\Delta_{1r}u(x,\cdot)$ is continuous.

To show that $\Delta_{1r}u(x,\cdot)$ satisfies the mean value equality, take $\tilde{y}_0 \in B^n(y_0,r_2)$ and $\rho > 0$ arbitrarily such that $\overline{B^n(\tilde{y}_0,\rho)} \subset B^n(y_0,r_2)$. Because of (6.1) we can use Fubini's theorem. Thus

$$\frac{1}{\nu_n\rho^n}\int\limits_{B^n(\tilde{y}_0,\rho)} \Delta_{1r}u(x,y)dm_n(y) =$$

$$=\frac{1}{\nu_n\rho^n}\int\limits_{B^n(\tilde{y}_0,\rho)} \{\frac{2(m+2)}{r^2}\cdot\frac{1}{\nu_m r^m}\int\limits_{B^m(0,r)} \big[u(x+x',y)-u(x,y)\big]dm_m(x')\}dm_n(y) =$$

$$=\frac{2(m+2)}{r^2}\cdot\frac{1}{\nu_m r^m}\int\limits_{B^m(0,r)} \{\frac{1}{\nu_n\rho^n}\int\limits_{B^n(\tilde{y}_0,\rho)} \big[u(x+x',y)-u(x,y)\big]dm_n(y)\}dm_m(x') =$$

$$=\frac{2(m+2)}{r^2}\cdot\frac{1}{\nu_m r^m}\int\limits_{B^m(0,r)} \big[u(x+x',\tilde{y}_0)-u(x,\tilde{y}_0)\big]\,dm_m(x') = \Delta_{1r}u(x,\tilde{y}_0).$$

Step 3 *The functions $\Delta_1 u(x,\cdot) : B^n(y_0,r_2) \to \mathbb{R}$, $x \in G\cap A$, are defined, nonnegative and harmonic. Here*

$$A := \bigcap_{k=1}^{+\infty} A(y_k),$$

where $H_2 = \{y_k, k = 1,2,\ldots\}$ (we may clearly suppose that H_2 is countable), and, for arbitrary $y \in B^n(y_0,r_2)$,

$$A(y) := \{x \in B^m(x_0,r_1) : \Delta_{1*}u(x,y) = \Delta_1^* u(x,y) = \Delta_1 u(x,y) \in \mathbb{R}\}.$$

The result [126], Lemma 2.2, p. 280 (see also [131], p. 451, and [133], p. 376) states that $m_m(B^m(x_0,r_1)\setminus A(y)) = 0$ for each $y \in B^n(y_0,r_2)$.

Take $x \in G\cap A$ and a sequence $r_j \to 0$, $0 < r_j < r_x$, $j = 1,2,\ldots$, arbitrarily. [49], Corollary 3 a), p. 6 (or [4], Lemma 1.5.6, p. 16) gives that the family

$$\Delta_{1r_j}u(x,\cdot) : B^n(y_0,r_2) \to \mathbb{R},\ j = 1,2,\ldots,$$

of nonnegative and harmonic functions is either uniformly equicontinuous and locally uniformly bounded, or else

$$\sup_{j=1,2,\ldots} \Delta_{1r_j}u(x,\cdot) \equiv +\infty.$$

On the other hand, since $x \in G\cap A$, we know that for each $y_k \in H_2$, $k = 1,2,\ldots$,

$$\Delta_{1r_j}u(x,y_k) \to \Delta_1 u(x,y_k) \in \mathbb{R}$$

as $j \to +\infty$. Therefore, by [153], Theorem 20.3, p. 68, and by [49], c), p. 2 (or [4], Theorem 1.5.8, p. 17), the limit

$$\Delta_1 u(x,\cdot) = \lim_{j\to+\infty} \Delta_{1r_j}u(x,\cdot)$$

exists and defines a harmonic function in $B^n(y_0,r_2)$. Since the limit is clearly independent of the considered sequence r_j, the claim follows.

Step 4 *The function $\Delta_1 u(\cdot,\cdot)\mid (G\cap H_1\cap A\cap B)\times B^n(y_0,r_2)$ has a continuous extension $\tilde{\Delta}_1 u(\cdot,\cdot):(A\cap B)\times B^n(y_0,r_2)\to\mathbb{R}$. Moreover, the functions $\tilde{\Delta}_1 u(x,\cdot):B^n(y_0,r_2)\to\mathbb{R}$, $x\in A\cap B$, are nonnegative and harmonic. Here*

$$B:=\bigcap_{k=1}^{+\infty} B(y_k),$$

where, for arbitrary $y\in B^n(y_0,r_2)$, we use the notation

$$B(y):=\{x\in B^m(x_0,r_1):\Delta_1 u(x',y)\to\Delta_1 u(x,y) \text{ as } x'\to x,\ x'\in H_1\}.$$

Using the assumption (d1), one sees easily that $G\cap H_1\cap A\cap B$ is dense in $A\cap B$.

To show the existence of the desired continuous extension, it is clearly sufficient to show that for each $(\tilde{x}_0,\tilde{y}_0)\in(A\cap B)\times B^n(y_0,r_2)$, the limit

$$\lim_{(x,y)\to(\tilde{x}_0,\tilde{y}_0),\,(x,y)\in(G\cap H_1\cap A\cap B)\times B^n(y_0,r_2)}\Delta_1 u(x,y)$$

exists. (This is of course standard, see e.g. [23], (3.15.5), p. 54.) To see this, it is sufficient to show that, for an arbitrary sequence $(x_j,y_j)\to(\tilde{x}_0,\tilde{y}_0)$, $(x_j,y_j)\in(G\cap H_1\cap A\cap B)\times B^n(y_0,r_2)$, $j=1,2,\ldots$, the limit

$$\lim_{j\to+\infty}\Delta_1 u(x_j,y_j)$$

exists.

That this limit indeed exists, is seen as above, just using the facts:

- the functions $\Delta_1 u(x_j,\cdot)$, $j=1,2,\ldots$, are nonnegative and harmonic in $B^n(y_0,r_2)$, by Step 3;
- for each $y_k\in H_2$, $k=1,2,\ldots$, $\Delta_1 u(x_j,y_k)\to\Delta_1 u(\tilde{x}_0,y_k)\in\mathbb{R}$ as $j\to+\infty$.

(See again [49], Corollary 3 a), p. 6 (or [4], Lemma 1.5.6, p. 16) and [153], Theorem 20.3, p. 68). That the functions $\tilde{\Delta}_1 u(x,\cdot):B^n(y_0,r_2)\to\mathbb{R}$, $x\in A\cap B$, are harmonic, see [49], c), p. 2 (or [4], Theorem 1.5.8, p. 17).

Step 5 *For each $x\in B^m(x_0,r_1)$ the functions*

$$B^n(y_0,r_2)\ni y\mapsto\tilde{v}(x,y):=\int G_{B^m(x_0,r_1)}(x,z)\tilde{\Delta}_1 u(z,y)dm_m(z)\in\mathbb{R}$$

and

$$B^n(y_0,r_2)\ni y\mapsto\tilde{h}(x,y):=u(x,y)+\tilde{v}(x,y)\in\mathbb{R}$$

are harmonic. Above and below $G_{B^m(x_0,r_1)}(x,z)$ is the Green function of the ball $B^m(x_0,r_1)$, with x as a pole.

Using Fubini's theorem one sees easily that for each $x\in B^m(x_0,r_1)$ the function $\tilde{v}(x,\cdot)$ satisfies the mean value equality. To see that $\tilde{v}(x,\cdot)$ is harmonic, it is sufficient to show that $\tilde{v}(x,\cdot)\in\mathcal{L}^1_{\mathrm{loc}}(B^n(y_0,r_2))$. Using just Fatou's lemma, one sees that $\tilde{v}(x,\cdot)$ is lower semicontinuous, hence superharmonic. Therefore either $\tilde{v}(x,\cdot)\equiv+\infty$ or else $\tilde{v}(x,\cdot)\in\mathcal{L}^1_{\mathrm{loc}}(B^n(y_0,r_2))$. The following argument shows that the former alternative cannot occur. Indeed, for each $x \in A \cap B$ and for each $y_k \in H_2$, $k = 1, 2, \ldots$, we see, using the definition of the (continuous) function $\tilde{\Delta}_1 u(\cdot,\cdot)$ and (d1), that

$$\begin{aligned}\tilde{\Delta}_1 u(x, y_k) &= \lim_{x' \to x, x' \in G \cap H_1 \cap A \cap B} \tilde{\Delta}_1 u(x', y_k) = \\ &= \lim_{x' \to x, x' \in G \cap H_1 \cap A \cap B} \Delta_1 u(x', y_k) = \Delta_1 u(x, y_k) \in \mathbb{R}.\end{aligned}$$

Hence $\tilde{v}(x, y_k) = v(x, y_k) \in \mathbb{R}$ for each $x \in B^m(x_0, r_1)$ and $y_k \in H_2$, $k = 1, 2, \ldots$. (See (6.2) below in Step 6 for the definition of $v(\cdot,\cdot) : B^m(x_0, r_1) \times B^n(y_0, r_2) \to \mathbb{R}$.) Therefore, for each $x \in B^m(x_0, r_1)$, the function $\tilde{v}(x, \cdot)$ and thus also the function $\tilde{h}(x, \cdot) = u(x, \cdot) + \tilde{v}(x, \cdot)$ are harmonic.

Step 6 *For each $y \in B^n(y_0, r_2)$ the function*

$$B^m(x_0, r_1) \ni x \mapsto \tilde{h}(x, y) \in \mathbb{R}$$

is harmonic.

With the aid of the version of the Riesz Decomposition Theorem, given in [126], 1.3. Theorem II, p. 279, and p. 278, too (see also [132], Theorem 1, p. 499), for each $y \in B^n(y_0, r_2)$ one can write

$$u(x, y) = h(x, y) - v(x, y),$$

where

$$v(x, y) := \int G_{B^m(x_0, r_1)}(x, z) \Delta_1 u(z, y) dm_m(z) \tag{6.2}$$

and $h(\cdot, y)$ is the least harmonic majorant of $u(\cdot, y) \mid B^m(x_0, r_1)$. Here $v(\cdot, y)$ is continuous and superharmonic in $B^m(x_0, r_1)$.

As shown above in (6.2), $v(\cdot, y_k) = \tilde{v}(\cdot, y_k)$ for each $y_k \in H_2$, $k = 1, 2, \ldots$. Therefore $\tilde{h}(\cdot, y_k) = h(\cdot, y_k)$, and thus $\tilde{h}(\cdot, y_k)$ is harmonic for each $y_k \in H_2$, $k = 1, 2, \ldots$.

To see that $\tilde{h}(\cdot, y)$ is harmonic also for $y \in B^n(y_0, r_2) \setminus H_2$, take $\tilde{y}_0 \in B^n(y_0, r_2) \setminus H_2$ arbitrarily, and proceed in the following way. Take $z \in A \cap B \cap A(\tilde{y}_0) \cap C(\tilde{y}_0)$ arbitrarily, where, for arbitrary $y \in B^n(y_0, r_2) \setminus H_2$,

$$C(y) := \{z \in B^m(x_0, r_1) : \Delta_{1*} u(z, y') \to \Delta_{1*} u(z, y) \text{ as } y' \to y,\ y' \in H_2\}.$$

Since $z \in A(\tilde{y}_0)$, we have $\Delta_{1*} u(z, \tilde{y}_0) = \Delta_1 u(z, \tilde{y}_0) \in \mathbb{R}$. Thus we may also suppose that $\Delta_{1*} u(z, y') = \Delta_1 u(z, y') \in \mathbb{R}$. Using then our assumption (d2) and the continuity of $\tilde{\Delta}_1 u(\cdot,\cdot)$, we see that

$$\Delta_1 u(z, \tilde{y}_0) = \tilde{\Delta}_1 u(z, \tilde{y}_0)$$

for every $z \in A \cap B \cap A(\tilde{y}_0) \cap C(\tilde{y}_0)$. Therefore, $\tilde{v}(x, \tilde{y}_0) = v(x, \tilde{y}_0)$ and thus $\tilde{h}(x, \tilde{y}_0) = h(x, \tilde{y}_0)$ for each $x \in B^m(x_0, r_1)$.

Step 7 *The use of the results of Lelong and of Avanissian.*

By Steps 5 and 6 we know that $\tilde{h}(\cdot,\cdot) = h(\cdot,\cdot)$ is separately harmonic in $B^m(x_0, r_1) \times B^n(y_0, r_2)$. Lelong's result [66], p. 561 (see also [7], Théorème 1, pp. 4–5) states that $\tilde{h}(\cdot,\cdot)$ is harmonic and thus locally bounded above in $B^m(x_0, r_1) \times B^n(y_0, r_2)$. Therefore also $u(\cdot,\cdot)$ is locally bounded above in $B^m(x_0, r_1) \times B^n(y_0, r_2)$. But then it follows from [6], Théorème 9, p. 140 (or of course from any of the already cited newer results of Arsove, ours, Cegrell's and Sadullaev's, and Armitage's and Gardiner's) that $u(\cdot,\cdot)$ is subharmonic on $B^m(x_0,r_1)\times B^n(y_0,r_2)$. □

Remark 6.6. Observe that the assumption (d2) was needed *only* to see that

$\Delta_1 u(x,y)=\tilde{\Delta}_1 u(x,y)$ for almost every $x\in B^m(x_0,r_1)$ and for each $y\in B^n(y_0,r_2)\setminus H_2$.

(At this point one might recall that the functions $\tilde{\Delta}_1 u(x,\cdot): B^n(y_0,r_2)\to\mathbb{R}$, $x\in B^m(x_0,r_1)$, are harmonic.)

From the above proof one sees easily that the assumption (d), that is (d1) and (d2), can be replaced by:

(d*) *There is a set* $H_1^*\subset B^m(x_0,r_1)$, *dense in* $B^m(x_0,r_1)$, *such that for each* $y\in B^n(y_0,r_2)$, *for almost every* $x\in B^m(x_0,r_1)$ *and for each* $x\in H_1^*$,

$$\Delta_1 u(x',y)\to\Delta_1 u(x,y)\in\mathbb{R}$$

as $x'\to x$, $x'\in H_1^*$.

Though our Theorem 6.5 might be considered somewhat technical, it has, nevertheless, the following concise corollaries, which both already improve the result of Kołodziej and Thorbiörnson, Theorem 6.1 above. The first corollary improves our previous result [109], Corollary 5.1, p. 74, [113], Corollary 4.3.3, p. e2626, the second one has been given before.

Corollary 6.7. *Let* Ω *be a domain in* $\mathbb{R}^{m+n}$, $m,n\geq 2$. *Let* $u:\Omega\to\mathbb{R}$ *be such that*

(a) *for each* $y\in\mathbb{R}^n$ *the function*

$$\Omega(y)\ni x\mapsto u(x,y)\in\mathbb{R}$$

is continuous and subharmonic,

(b) *for each* $x\in\mathbb{R}^m$ *the function*

$$\Omega(x)\ni y\mapsto u(x,y)\in\mathbb{R}$$

is harmonic,

(c) *for each* $y\in\mathbb{R}^n$ *one has* $\Delta_{1*}u(x,y)<+\infty$ *for each* $x\in\Omega(y)$, *possibly with the exception of a polar set in* $\Omega(y)$,

(d) *there is a set* $H_3\subset\mathbb{R}^m$ *dense in* $\mathbb{R}^m$ *such that for each* $y\in\mathbb{R}^n$ *and for almost every* $x\in\Omega(y)$ *and for each* $x\in H_3$,

$$\Delta_1 u(x',y)\to\Delta_1 u(x,y)\in\mathbb{R}$$

as $x'\to x$, $x'\in H_3$.

Then u *is subharmonic in* Ω.

Corollary 6.8. ([107], Theorem 6, p. 234, and [108], Corollary, p. 444) *Let* Ω *be a domain in* $\mathbb{R}^{m+n}$, $m,n\geq 2$. *Let* $u:\Omega\to\mathbb{R}$ *be such that*

(a) *for each* $y\in\mathbb{R}^n$ *the function*

$$\Omega(y)\ni x\mapsto u(x,y)\in\mathbb{R}$$

is continuous and subharmonic,

(b) *for each* $x \in \mathbb{R}^m$ *the function*

$$\Omega(x) \ni y \mapsto u(x,y) \in \mathbb{R}$$

is harmonic,

(c) *for each* $y \in \mathbb{R}^n$ *the function*

$$\Omega(y) \ni x \mapsto \Delta_1 u(x,y) \in \mathbb{R}$$

is defined and continuous.

Then u *is subharmonic in* Ω.

7. Weighted boundary behavior of quasinearly subharmonic functions

Abstract. We give certain weighted boundary behavior properties for quasinearly subharmonic functions, related to previous results of Gehring, Hallenbeck, Mizuta, Pavlović, Stoll, Suzuki and others. Especially, we give a limiting case result of a nonintegrability result of Suzuki.

Keywords. Subharmonic, quasinearly subharmonic, boundary behavior, approach region, radial order of a subharmonic function, Ahlfors-regular set, weighted nonintegrability

7.1. **Admissible functions and approach regions.** A function $\varphi : [0,+\infty) \to [0,+\infty)$ is *admissible*, if it is strictly increasing, surjective, and there are constants $C_0 = C_0(\varphi) \geq 1$ and $r_0 > 0$ such that

$$\varphi(2t) \leq C_0\,\varphi(t) \quad \text{and} \quad \varphi^{-1}(2s) \leq C_0\,\varphi^{-1}(s) \tag{7.1}$$

for all s, t, $0 \leq s, t \leq r_0$.

The assumptions that admissible functions are surjective and globally strictly increasing, are not fully needed in the proof of our Theorem 7.5 below. We have, however, preferred to make these assumptions for the sake of simplicity. As a matter of fact, with the aid of these assumptions we can avoid certain, perhaps slightly technical measurability considerations and also some (local) existence questions of the inverse of an admissible function. The same remark applies, at least partly, also to the above definition of permissible functions.

Nonnegative, increasing surjective functions $\varphi_1(t)$, which satisfy the Δ_2-condition and for which the functions $t \mapsto \frac{\varphi_1(t)}{t}$ are increasing, are examples of admissible functions. Further examples are $\varphi_2(t) = ct^{\alpha}[\log(\delta + t^{\gamma})]^{\beta}$, where $c > 0$, $\alpha > 0$, $\delta \geq 1$, and $\beta, \gamma \in \mathbb{R}$ are such that $\alpha + \beta\gamma > 0$.

Let $\varphi : [0,+\infty) \to [0,+\infty)$ be an admissible function and let $\alpha > 0$. One says that $\zeta \in \partial D$ is (φ,α)-*accessible*, shortly *accessible*, if

$$\Gamma_{\varphi}(\zeta,\alpha) \cap B^N(\zeta,\rho) \neq \emptyset$$

for all $\rho > 0$. Here

$$\Gamma_{\varphi}(\zeta,\alpha) = \{\, x \in D : \varphi(|x-\zeta|) < \alpha\,\delta(x) \,\},$$

and it is called a (φ,α)-*approach region*, shortly an *approach region, in D at* ζ.

Mizuta [73] has considered boundary limits of harmonic functions in Sobolev-Orlicz classes on bounded Lipschitz domains U of $\mathbb{R}^N$, $N \geq 2$. His approach regions are of the form

$$\Gamma_{\phi}(\zeta,\alpha) = \{\, x \in U : \phi(|x-\zeta|) < \alpha\,\delta(x) \,\},$$

where now, instead of (7.1), $\phi : [0,+\infty) \to [0,+\infty)$ is an increasing function which satisfies the Δ_2-condition and is such that $t \mapsto \frac{\phi(t)}{t}$ is increasing. As pointed out above, such functions are related to our admissible functions. In fact, they form a proper subclass of our admissible functions.

Juhani Riihentaus

7.2. **Previous weighted boundary behavior results of Gehring, Hallenbeck, Stoll, Mizuta and ours.** The classical result on the area is due to Gehring [39], Theorem 1, p. 77, and Hallenbeck [43], Theorems 1 and 2, pp. 117-118. Gehring considered the case $p > 1$, and his proof was based on a theorem of Hardy-Littlewood. Much later Hallenbeck proved the general case $p > 0$ with the aid of a generalized mean value inequality (1.2) for subharmonic functions.

Theorem 7.1. *Suppose w is a nonnegative subharmonic function in the unit disc $\mathbb{D} = \{ z \in \mathbb{C} : |z| < 1 \}$ satisfying*

$$\iint_{\mathbb{D}} w(z)^p dxdy < +\infty, \quad z = x + iy,$$

for some $p > 0$. Then for almost every θ,

$$\lim_{r \to 1-} \sup_{z \in \Gamma_\alpha(e^{i\theta}), |z| \geq r} (1 - |z|) w(z)^p = 0.$$

Here

$$\Gamma_\alpha(e^{i\theta}) = \{ z \in \mathbb{D} : |e^{i\theta} - z| < \alpha(1 - |z|) \}.$$

Stoll [140], Theorem 2, p. 307, gave the following improvement:

Theorem 7.2. *Let f be a nonnegative subharmonic function on a domain G in $\mathbb{R}^N$, $G \neq \mathbb{R}^N$, $N \geq 2$, with $\mathcal{C}^1$ boundary. Let*

$$\int_G \delta(x)^\gamma f(x)^p \, dm_N(x) < +\infty \tag{7.2}$$

for some $p > 0$ and $\gamma > -1 - \beta(p)$, where $\beta(p) := \max\{ (N-1)(1-p), 0 \}$. Let $0 < d \leq N - 1$. Then for each $\tau \geq 1$ and $\alpha > 0$ ($\alpha > 1$ when $\tau = 1$), there exists a subset E_τ of ∂G with $\mathcal{H}^d(E_\tau) = 0$ such that

$$\lim_{\rho \to 0} \{ \sup_{x \in \Gamma_{\tau,\alpha,\rho}(\zeta)} [\delta(x)^{N+\gamma-\frac{d}{\tau}} f(x)^p] \} = 0$$

for all $\zeta \in \partial G \setminus E_\tau$. Above, for $\zeta \in \partial G$ and $\rho > 0$,

$$\Gamma_{\tau,\alpha,\rho}(\zeta) = \Gamma_{\tau,\alpha}(\zeta) \cap G_\rho,$$

where

$$\Gamma_{\tau,\alpha}(\zeta) = \{ x \in G : |x - \zeta|^\tau < \alpha\delta(x) \}, \quad G_\rho = \{ x \in G : \delta(x) < \rho \}.$$

Remark 7.3. Observe that Suzuki [145], Theorem 2, p. 271, has shown that if $p > 0$ and $\gamma \leq -1 - \beta(p)$, then the only nonnegative subharmonic function on a bounded domain D with $\mathcal{C}^2$ (as a matter of fact, for $\mathcal{C}^{1,1}$) boundary satisfying (7.2) is the zero function. On the other hand, again by [145], pp. 272-273, if $p > 0$ and $\gamma > -1 - \beta(p)$, then there exist nonnegative non-zero subharmonic functions in B^N satisfying (7.2).

In [97], Theorem, p. 233, we improved Stoll's result for rather general domains and more general approach regions. Using a different method, Mizuta [75], Theorem 2, p. 73, gave the following related result:

Theorem 7.4. *Let u be a nonnegative quasinearly subharmonic function on a bounded domain D in $\mathbb{R}^N$, $N \geq 2$, satisfying the condition*

$$\int_D \delta(x)^\gamma u(x)^p \, dm_N(x) < +\infty$$

for some $p > 0$ and some $\gamma \in \mathbb{R}$. Then for $\tau \geq 1$, $\alpha > 0$, and d, $0 \leq d < N$, there exists a set $A_{\tau,d} \subset \partial D$ such that $\mathcal{H}^d(A_{\tau,d}) = 0$, and

$$\lim_{x \to \zeta,\, x \in \mathcal{S}_{\tau,\alpha}(\zeta)} \delta(x)^{N+\gamma-\frac{d}{\tau}} u(x)^p = 0$$

for every $\zeta \in \partial D \setminus A_{\tau,d}$. Here

$$\mathcal{S}_{\tau,\alpha}(\zeta) = \{x \in D : |\, x - \zeta\,|^\tau < \alpha\delta(x)\}.$$

Observe that though Mizuta does not explicitly consider accessible points $\zeta \in \partial D$, his conclusion is of course meaningful only for such points. (We have formulated here Mizuta's result using partly our above terminology and notation; Mizuta's constant $\delta > 0$, say, is now $(N + \gamma - \frac{d}{\tau})/p$.)

The following result [105], Theorem, p. 31, or essentially equivalently [85], Theorem 4, p. 102, includes and improves the above Theorem 7.1, Theorem 7.2, Theorem 7.4, and also our previous result [97], Theorem, p. 233. Observe that our result below covers now, in addition to the case of an arbitrary domain D of $\mathbb{R}^N$, $D \neq \mathbb{R}^N$, also all the cases $0 \leq d \leq N$. The proof of our result follows the lines of [97], proof of Theorem, pp. 235-238, however, with some additional elements. It is essentially different from Mizuta's proof.

Theorem 7.5. *Suppose that u is a nonnegative quasinearly subharmonic function on D. Let $\varphi : [0, +\infty) \to [0, +\infty)$ be an admissible function and $\alpha > 0$. Suppose that*

$$\int_D \delta(x)^\gamma u(x) \, dm_N(x) < +\infty$$

for some $\gamma \in \mathbb{R}$. Then for each d, $0 \leq d \leq N$, there is a set $E_d \subset \partial D$ such that $\mathcal{H}^d(E_d) = 0$ and

$$\lim_{\rho \to 0} \Big(\sup_{x \in \Gamma_{\varphi,\rho}(\zeta,\alpha)} \{\delta(x)^{N+\gamma} [\varphi^{-1}(\delta(x))]^{-d} u(x)\} \Big) = 0$$

for every (φ, α)-accessible point $\zeta \in \partial D \setminus E_d$. Here

$$\Gamma_{\varphi,\rho}(\zeta,\alpha) = \{x \in D : \varphi(|\, x - \zeta\,|) < \alpha\delta(x),\ \delta(x) < \rho\}.$$

For the proof of our result we need four lemmas.

Lemma 7.6. ([97], Lemma 2.2, p. 234) *Let $\zeta \in \partial D$. Then $B(x) \subset \Gamma_{\varphi,\rho'}(\zeta, \alpha')$ for all $x \in \Gamma_{\varphi,\rho}(\zeta, \alpha)$, where $\alpha' = \alpha C_0$, $0 < \rho \leq \frac{\varphi(r_0)}{\alpha}$ and $\rho' = \frac{4}{3}\rho$.*

Above (and in the sequel) $C_0 = C_0(\varphi)$ and r_0 are the constants involved in the definition for the considered admissible function φ, see above (7.1). Recall that $B(x) = B^N(x, \frac{1}{3}\delta(x))$.

Write for $x \in D$,

$$\widetilde{\Gamma}_\varphi(x,\alpha) := \{\xi \in \partial D : x \in \Gamma_\varphi(\xi,\alpha)\}.$$

Lemma 7.7. ([97], Lemma 2.3, p. 234) *There are constants $C_1 = C_1(C_0,\alpha) \geq 1$, $C_2 = C_2(C_0,\alpha) \geq 1$ and $C_3 = C_3(C_0,\alpha) \geq 1$ such that for any $\zeta \in \partial D$, any ρ, $0 < \rho \leq \min\{\frac{r_0}{2^{1+\alpha}}, \frac{r_0}{2^{3\alpha C_0}}, \frac{\varphi(r_0)}{\alpha}\}$, and any $x_0 \in \Gamma_{\varphi,\rho}(\zeta,\alpha)$,*

$$\begin{aligned}\widetilde{\Gamma}_\varphi(x,\alpha) \subset & B^N(x, C_1\varphi^{-1}(\delta(x))) \subset \\ & \subset B^N(x_0, C_1C_2\varphi^{-1}(\delta(x_0))) \subset B^N(\zeta, C_1C_2C_3\varphi^{-1}(\delta(x_0)))\end{aligned}$$

for all $x \in B(x_0)$.

Proof. One sees easily that $C_1 = C_0^{1+\alpha}$ works in the first inclusion. We prove below the second inclusion and leave the third to the reader.

Observe first that for all $x' \in B(x_0)$, one has $\frac{2}{3}\delta(x_0) \leq \delta(x') \leq \frac{4}{3}\delta(x_0)$. Thus

$$\varphi^{-1}(\delta(x')) \leq \varphi^{-1}(2\delta(x_0)) \leq C_0\varphi^{-1}(\delta(x_0)). \tag{7.3}$$

Take $z \in B^N(x, C_1\varphi^{-1}(\delta(x)))$ arbitrarily. Thus by (7.3),

$$|z - x| < C_1\varphi^{-1}(\delta(x)) \leq C_0C_1\varphi^{-1}(\delta(x_0)).$$

We get the claim observing first that

$$|z - x_0| \leq |z - x| + |x - \zeta| + |\zeta - x_0|,$$

and using the following estimates. Since $x_0 \in \Gamma_\varphi(\zeta,\alpha)$, we have

$$|\zeta - x_0| < \varphi^{-1}(\alpha\delta(x_0)) \leq C_0^{1+\alpha}\varphi^{-1}(\delta(x_0)).$$

By Lemma 7.6, we know that $x \in \Gamma_\varphi(\zeta,\alpha')$. Thus using again also (7.3),

$$|x - \zeta| < \varphi^{-1}(\alpha'\delta(x)) \leq C_0^{1+\alpha'}\varphi^{-1}(\delta(x)) \leq C_0^{2+\alpha'}\varphi^{-1}(\delta(x_0)) = C_0^{2+\alpha C_0}\varphi^{-1}(\delta(x_0)).$$

Combined we have

$$|z - x_0| < (C_0C_1 + C_0^{1+\alpha} + C_0^{2+\frac{3}{2}\alpha C_0})\varphi^{-1}(\delta(x_0)).$$

Hence

$$B^N(x, C_1\varphi^{-1}(\delta(x))) \subset B^N(x_0, C_1C_2\varphi^{-1}(\delta(x_0)))$$

with $C_2 = 1 + C_0 + C_0^{1-\alpha+\frac{3}{2}\alpha C_0}$. Above one has needed the assumption that ρ is small enough, say $0 < \rho \leq \frac{r_0}{2^{3\alpha C_0}}$, in the case of this inclusion. □

Lemma 7.8. ([71], Theorem 6.2 (1), p. 89) *Suppose $A \subset \mathbb{R}^N$ with $\mathcal{H}^d(A) < +\infty$, $0 \leq d \leq N$. Then*

$$\limsup_{r \to 0} \frac{\mathcal{H}^d(A \cap B^N(x,r))}{(2r)^d} \leq 1$$

for $\mathcal{H}^d$-almost all $x \in A$.

Lemma 7.9. ([71], Theorem 8.13, p. 117) *Let $0 \leq d \leq N$. For any Borel set $B \subset \mathbb{R}^N$,*

$$\mathcal{H}^d(B) = \sup\{\mathcal{H}^d(C) : C \subset B \text{ is compact with } \mathcal{H}^d(C) < +\infty\}.$$

Proof. We begin with the proof of Theorem 7.5. Write

$$E=\{\zeta\in\partial D:\lim_{\rho\to 0}(\sup_{x\in\Gamma_{\varphi,\rho}(\zeta,\alpha)}\{\delta(x)^{N+\gamma}[\varphi^{-1}(\delta(x))]^{-d}u(x)\})>0\}.$$

We must show that $\mathcal{H}^d(E)=0$. Observe first that E is a Borel set. This follows at once from the fact that, for each $\rho>0$ the function $M_\rho:\partial D\to[0,+\infty]$,

$$M_\rho(\zeta)=\sup_{x\in\Gamma_{\varphi,\rho}(\zeta,\alpha)}\{\delta(x)^{N+\gamma}[\varphi^{-1}(\delta(x))]^{-d}u(x)\},$$

is lower semicontinuous.

By Lemma 7.9 it is sufficient to show that $\mathcal{H}^d(K)=0$ if $K\subset E$ is compact and $\mathcal{H}^d(K)<+\infty$.

Though the rest of the proof goes along the same lines as in [97], proof of Theorem, pp. 235-238, we write the details down, just for the convenience of the reader.

For every $\zeta\in\partial D$ and for every ρ sufficiently small, say, $0<\rho\leq\min\{\frac{r_0}{2^{1+\alpha}},\frac{r_0}{2^{3\alpha C_0}},\frac{\varphi(r_0)}{\alpha}\}$, we write

$$M_\rho^K(\zeta)=\sup_{x\in\Gamma_{\varphi,\rho}(\zeta,\alpha)}\frac{\delta(x)^{N+\gamma}u(x)}{[\varphi^{-1}(\delta(x))]^d+\mathcal{H}^d(B^N(x,C_1C_2\varphi^{-1}(\delta(x)))\cap K)},$$

$$M^K(\zeta)=\liminf_{\rho\to 0}M_\rho^K(\zeta).$$

If $\Gamma_{\varphi,\rho}(\zeta,\alpha)=\emptyset$, then define $M_\rho^K(\zeta)=0$. Above and in the sequel C_1 and C_2 (and C_3) are the constants from Lemma 7.6. Clearly M_ρ^K and M^K are Borel functions on K. (M_ρ^K is even lower semicontinuous; recall that $\varphi:[0,+\infty)\to[0,+\infty)$ is an increasing surjection, thus continuous.)

Suppose for a while that $\zeta\in K$ and $\rho>0$ are fixed. Take $x_0\in\Gamma_{\varphi,\rho}(\zeta,\alpha)$ arbitrarily. Using Lemma 7.6 (the second inclusion) and (7.3) above, one gets

$$\frac{\delta(x_0)^{N+\gamma}u(x_0)}{[\varphi^{-1}(\delta(x_0))]^d+\mathcal{H}^d(B^N(x_0,C_1C_2\varphi^{-1}(\delta(x_0)))\cap K)}\leq$$
$$\leq C\int_{B(x_0)}\frac{\delta(x)^{\gamma}u(x)}{[\varphi^{-1}(\delta(x))]^d+\mathcal{H}^d(B^N(x,C_1\varphi^{-1}(\delta(x)))\cap K)}dm_N(x).$$

Taking supremum on the left over $x_0\in\Gamma_{\varphi,\rho}(\zeta,\alpha)$, and using the fact that, by Lemma 7.6, $B(x_0)\subset\Gamma_{\varphi,\rho'}(\zeta,\alpha')$ for all such x_0, we get

$$M_\rho^K(\zeta)\leq C\int_{\Gamma_{\varphi,\rho'}(\zeta,\alpha')}\frac{\delta(x)^{\gamma}u(x)}{[\varphi^{-1}(\delta(x))]^d+\mathcal{H}^d(B^N(x,C_1\varphi^{-1}(\delta(x)))\cap K)}dm_N(x)\leq$$
$$\leq C\int_{D_{\rho'}}\chi_{\Gamma_\varphi(\zeta,\alpha')}(x)\frac{\delta(x)^{\gamma}u(x)}{[\varphi^{-1}(\delta(x))]^d+\mathcal{H}^d(B^N(x,C_1\varphi^{-1}(\delta(x)))\cap K)}dm_N(x).$$

Recall that $\rho'=\frac{4}{3}\rho$ and $\alpha'=\frac{3}{2}\alpha C_0$. Next integrate on both sides of the last inequality over $\zeta\in K$, with respect to the Hausdorff measure $\mathcal{H}^d$, use Fubini's theorem and

Lemma 7.7 (the first inclusion):

$$\int_K M_\rho^K(\zeta)\,d\mathcal{H}^d(\zeta) \leq$$

$$\leq C\int_K \{\int_{D_{\rho'}} \chi_{\Gamma_\varphi(\zeta,\alpha')}(x)\frac{\delta(x)^\gamma u(x)}{[\varphi^{-1}(\delta(x))]^d + \mathcal{H}^d(B^N(x,C_1\varphi^{-1}(\delta(x)))\cap K)}\,dm_N(x)\}\,d\mathcal{H}^d(\zeta) \leq$$

$$\leq C\int_{D_{\rho'}} [\int_K \chi_{\Gamma_\varphi(\zeta,\alpha')}(x)\,d\mathcal{H}^d(\zeta)]\frac{\delta(x)^\gamma u(x)}{[\varphi^{-1}(\delta(x))]^d + \mathcal{H}^d(B^N(x,C_1\varphi^{-1}(\delta(x)))\cap K)}\,dm_N(x) \leq$$

$$\leq C\int_{D_{\rho'}} [\int_K \chi_{\widetilde{\Gamma}_\varphi(x,\alpha')}(\zeta)\,d\mathcal{H}^d(\zeta)]\frac{\delta(x)^\gamma u(x)}{[\varphi^{-1}(\delta(x))]^d + \mathcal{H}^d(B^N(x,C_1\varphi^{-1}(\delta(x)))\cap K)}\,dm_N(x) \leq$$

$$\leq C\int_{D_{\rho'}} \mathcal{H}^d(\widetilde{\Gamma}_\varphi(x,\alpha')\cap K)\frac{\delta(x)^\gamma u(x)}{[\varphi^{-1}(\delta(x))]^d + \mathcal{H}^d(B^N(x,C_1\varphi^{-1}(\delta(x)))\cap K)}\,dm_N(x) \leq$$

$$\leq C\int_{D_{\rho'}} \delta(x)^\gamma u(x)\,dm_N(x).$$

Using Fatou's lemma, the last inequality above, and the assumption of the weighted integrability of u, we get

$$\int_K M^K(\zeta)\,d\mathcal{H}^d(\zeta) = \int_K [\lim_{\rho\to 0} M_\rho^K(\zeta)]\,d\mathcal{H}^d(\zeta) = \int_K [\liminf_{\rho\to 0} M_\rho^K(\zeta)]\,d\mathcal{H}^d(\zeta) \leq$$

$$\leq \liminf_{\rho\to 0}\int_K M_\rho^K(\zeta)\,d\mathcal{H}^d(\zeta) \leq \liminf_{\rho\to 0}[C\int_{D_{\rho'}} \delta(x)^\gamma u(x)\,dm_N(x)] \leq$$

$$\leq \lim_{\rho\to 0}[C\int_{D_{\rho'}} \delta(x)^\gamma u(x)\,dm_N(x)] = 0.$$

Therefore

$$M^K(\zeta) = 0 \text{ for } \mathcal{H}^d - \text{almost } \zeta \in K.$$

To complete the proof, suppose that $\zeta \in K$ is accessible and such that $M^K(\zeta) = 0$. Using the assumption $\mathcal{H}^d(K) < +\infty$ and Lemma 7.8, we may moreover suppose that

$$\limsup_{\rho\to 0}\frac{\mathcal{H}^d(B^N(\zeta,\rho)\cap K)}{(2\rho)^d} \leq 1.$$

Thus for all $x_0 \in \Gamma_{\varphi,\rho}(\zeta,\alpha)$ and ρ small enough,

$$\mathcal{H}^d(B^N(\zeta,C_1C_2C_3\varphi^{-1}(\delta(x_0)))\cap K) \leq 2^{d+1}\,(C_1C_2C_3)^d[\varphi^{-1}(\delta(x_0))]^d. \tag{7.4}$$

Using the notation

$$\widetilde{C} = \frac{1}{1+2^{d+1}(C_1C_2C_3)^d},$$

one gets, with the aid of (7.4) and using Lemma 7.7 (the third inclusion),

$$\widetilde{C}\lim_{\rho\to 0}\Big(\sup_{x_0\in\Gamma_{\varphi,\rho}(\zeta,\alpha)}\{\delta(x_0)^{N+\gamma}[\varphi^{-1}(\delta(x_0))]^{-d}u(x_0)\}\Big)\leq$$

$$\leq\lim_{\rho\to 0}\Big\{\sup_{x_0\in\Gamma_{\varphi,\rho}(\zeta,\alpha)}\frac{\delta(x_0)^{N+\gamma}u(x_0)}{[\varphi^{-1}(\delta(x_0))]^d+\mathcal{H}^d(B^N(\zeta,C_1C_2C_3\varphi^{-1}(\delta(x_0)))\cap K)}\Big\}\leq$$

$$\leq\lim_{\rho\to 0}\Big\{\sup_{x_0\in\Gamma_{\varphi,\rho}(\zeta,\alpha)}\frac{\delta(x_0)^{N+\gamma}u(x_0)}{[\varphi^{-1}(\delta(x_0))]^d+\mathcal{H}^d(B^N(x_0,C_1C_2\varphi^{-1}(\delta(x_0)))\cap K)}\Big\}=$$

$$=\lim_{\rho\to 0}M^K_\rho(\zeta)=M^K(\zeta)=0.$$

Since $C(C_0,\alpha,d)=1/\widetilde{C}>0$, the proof is complete. □

7.3. **An application: On the radial order of a subharmonic function.** The following theorem is a special case of the above results of Gehring, Hallenbeck, and Stoll, Theorems 7.1 and 7.2:

Theorem 7.10. *If u is a function harmonic in the unit disc $\mathbb{D}$ of $\mathbb{C}$ such that*

$$I(u):=\int_{\mathbb{D}}|u(z)|^p(1-|z|)^\beta dm_2(z)<+\infty, \tag{7.5}$$

where $p>0$, $\beta>-1$, then

$$\lim_{r\to 1-}|u(re^{i\theta})|^p(1-r)^{\beta+1}=0 \tag{7.6}$$

for almost all $\theta\in[0,2\pi)$.

Recall that Gehring, Hallenbeck and Stoll in fact considered subharmonic functions and that the limit in (7.6) was uniform in Stolz approach regions (in Stoll's result in even more general regions).

With the aid of the following Theorem 7.11, [83], Theorem 1, pp. 433-434, Pavlović showed that the convergence in (7.6) is dominated. At the same time he pointed out that whole Theorem 7.10 follows from his result:

Theorem 7.11. *If u is a function harmonic in $\mathbb{D}$ satisfying (7.5), where $p>0$, $\beta>-1$, then*

$$J(u):=\int_0^{2\pi}\sup_{0<r<1}|u(re^{i\theta})|^p(1-r)^{\beta+1}d\theta<+\infty.$$

Moreover, there is a constant $C=C_{p,\beta}$ such that $J(u)\leq CI(u)$.

With the aid of our Theorem 7.5, one can extend Theorem 7.11 considerably: Instead of absolute values of harmonic functions in the unit disk $\mathbb{D}$ of the complex plane $\mathbb{C}$ we will consider nonnegative quasinearly subharmonic functions defined on rather general domains of $\mathbb{R}^N$, $N\geq 2$. See Theorem 7.12, Theorem 7.13 and Corollary 7.14 below and [112], Theorem 1, Theorem 2 and Corollary, pp. 131-132, 134.

Recall first the definition of Ahlfors-regular sets. Let $0 \leq d \leq N$. A set $E \subset \mathbb{R}^N$ is *Ahlfors-regular with dimension d* if it is closed and there is a constant $C_4 > 0$ such that

$$C_4^{-1} r^d \leq \mathcal{H}^d(E \cap B^N(x,r)) \leq C_4 r^d$$

for all $x \in E$ and $r > 0$. The smallest constant C_4 is called the *regularity constant* for E. Simple examples of Ahlfors-regular sets include d-planes and d-dimensional Lipschitz graphs. Also certain Cantor sets and self-similar sets are Ahlfors-regular. For more details, see [21], pp. 9-10.

First a partial generalization to Pavlović's result, [83], Theorem 1, pp. 433-434, or Theorem 7.11 above. Observe that though the constant C below in (7.7) does depend on K', it is, nevertheless, otherwise independent of the (K'-)quasinearly subharmonic function u.

Theorem 7.12. *Let G be a domain in $\mathbb{R}^N$, $N \geq 2$, $G \neq \mathbb{R}^N$, such that its boundary ∂G is Ahlfors-regular of dimension d, $0 \leq d \leq N$. Let $u : G \to [0,+\infty)$ be a K'-quasinearly subharmonic function. Let $\varphi : [0,+\infty) \to [0,+\infty)$ be an admissible function, with constants r_0 and C_0. Let $\alpha > 0$ be arbitrary. Let $\rho_0 :=$ $\min\{r_0/2^{1+\alpha}, r_0/2^{3\alpha C_0}, \varphi(r_0)/\alpha\}$. Let $\gamma \in \mathbb{R}$ be such that*

$$\int_G \delta(x)^\gamma u(x)\, dm_N(x) < +\infty.$$

Then there is a constant $C = C(N,G,d,\varphi,\alpha,\gamma,K')$ such that for all ρ, $0 < \rho \leq \rho_0$,

$$\int_{\partial G} \sup_{x \in \Gamma_{\varphi,\rho}(\zeta,\alpha)} \{\delta(x)^{N+\gamma}[\varphi^{-1}(\delta(x))]^{-d} u(x)\}\, d\mathcal{H}^d(\zeta) \leq C \int_{G_{\rho'}} \delta(x)^\gamma u(x)\, dm_N(x), \tag{7.7}$$

where $\rho' = \frac{4}{3}\rho$ and

$$\Gamma_{\varphi,\rho}(\zeta,\alpha) = \{x \in \Gamma_\varphi(\zeta,\alpha) : \delta(x) < \rho\}.$$

Proof. Proceeding as in [105], proof of Theorem (with $\psi = id$), pp. 31-37, see also [97], proof of Theorem, pp. 235-238, and choosing $K = \partial G$, one obtains

$$\int_{\partial G} M_\rho^{\partial G}(\zeta)\, d\mathcal{H}^d(\zeta) \leq C \int_{G_{\rho'}} \delta(x)^\gamma u(x)\, dm_N(x),$$

where $\rho' = \frac{4}{3}\rho$ and $M_\rho^{\partial G} : \partial G \to [0,+\infty]$,

$$M_\rho^{\partial G}(\zeta) = \sup_{x \in \Gamma_{\varphi,\rho}(\zeta,\alpha)} \frac{\delta(x)^{N+\gamma} u(x)}{[\varphi^{-1}(\delta(x))]^d + \mathcal{H}^d(B^N(x, C_1 C_2 \varphi^{-1}(\delta(x))) \cap \partial G)}.$$

Here the constants $C_1 = C_1(C_0,\alpha)$, $C_2 = C_2(C_0,\alpha)$ and $C_3 = C_3(C_0,\alpha)$ are, as pointed out above, directly from [105], Lemma 2.3, pp. 32-33, or [97], proof of Lemma 2.3, pp. 234-235. By this lemma one has, for each $\zeta \in \partial G$ and for each $x \in \Gamma_{\varphi,\rho}(\zeta,\alpha)$, $B^N(x, C_1C_2\varphi^{-1}(\delta(x))) \subset B^N(\zeta, C_1C_2C_3\varphi^{-1}(\delta(x)))$. Since ∂G is Ahlfors-regular with dimension d, we have

$$\mathcal{H}^d(B^N(\zeta, C_1C_2C_3\varphi^{-1}(\delta(x))) \cap \partial G) \leq C_4[C_1C_2C_3\varphi^{-1}(\delta(x))]^d$$

where also C_4 is a fixed constant. Therefore

$$M_\rho^{\partial G}(\zeta) = \sup_{x\in\Gamma_{\varphi,\rho}(\zeta,\alpha)} \frac{\delta(x)^{N+\gamma}u(x)}{[\varphi^{-1}(\delta(x))]^d + \mathcal{H}^d(B^N(x,C_1C_2\,\varphi^{-1}(\delta(x)))\cap\partial G)} \geq$$

$$\geq \sup_{x\in\Gamma_{\varphi,\rho}(\zeta,\alpha)} \frac{\delta(x)^{N+\gamma}u(x)}{[\varphi^{-1}(\delta(x))]^d + \mathcal{H}^d(B^N(\zeta,C_1C_2C_3\,\varphi^{-1}(\delta(x)))\cap\partial G)} \geq$$

$$\geq \sup_{x\in\Gamma_{\varphi,\rho}(\zeta,\alpha)} \frac{\delta(x)^{N+\gamma}u(x)}{[\varphi^{-1}(\delta(x))]^d + C_4(C_1C_2C_3)^d[\varphi^{-1}(\delta(x))]^d} \geq$$

$$\geq \frac{1}{1+(C_1C_2C_3)^dC_4} \sup_{x\in\Gamma_{\varphi,\rho}(\zeta,\alpha)} \{\,\delta(x)^{N+\gamma}[\varphi^{-1}(\delta(x))]^{-d}u(x)\,\}.$$

Hence

$$\int_{\partial G} \sup_{x\in\Gamma_{\varphi,\rho}(\zeta,\alpha)} \{\,\delta(x)^{N+\gamma}[\varphi^{-1}(\delta(x))]^{-d}u(x)\,\}\,d\mathcal{H}^d(\zeta) \leq C\int_{G_{\rho'}} \delta(x)^\gamma u(x)\,dm_N(x),$$

concluding the proof. □

Theorem 7.12 seems to be useful in many situations. For example, with the aid of it one gets the following generalizations to Pavlović's result, [83], Theorem 1, pp. 433-435, or Theorem 7.11 above:

Theorem 7.13. *Let G, d, u, φ, α be as above in Theorem 7.12. Suppose moreover that $\mathcal{H}^d(\partial G) < +\infty$. Let $\gamma\in\mathbb{R}$. Then there is a constant $C = C(N,G,d,\varphi,\alpha,\gamma,K')$ such that*

$$\int_{\partial G} \sup_{x\in\Gamma_{\varphi}(\zeta,\alpha)} \{\delta(x)^{N+\gamma}[\varphi^{-1}(\delta(x))]^{-d}u(x)\}\,d\mathcal{H}^d(\zeta) \leq C\int_G \delta(x)^\gamma u(x)\,dm_N(x).$$

Proof. We may clearly assume that

$$\int_G \delta(x)^\gamma u(x)\,dm_N(x) < +\infty.$$

By Theorem 7.12,

$$\int_{\partial G} \sup_{x\in\Gamma_{\varphi,\rho}(\zeta,\alpha)} \{\,\delta(x)^{N+\gamma}[\varphi^{-1}(\delta(x))]^{-d}u(x)\,\}\,d\mathcal{H}^d(\zeta) \leq C\int_{G\rho'} \delta(x)^\gamma u(x)\,dm_N(x).$$

Write

$$\Gamma^c_{\varphi,\rho_0}(\zeta,\alpha) := \{x\in\Gamma_\varphi(\zeta,\alpha):\ \delta(x)\geq\rho_0\}.$$

Since

$$\sup_{x\in\Gamma_\varphi(\zeta,\alpha)} \{\,\delta(x)^{N+\gamma}[\varphi^{-1}(\delta(x))]^{-d}u(x)\,\} \leq \sup_{x\in\Gamma^c_{\varphi,\rho_0}(\zeta,\alpha)} \{\,\delta(x)^{N+\gamma}[\varphi^{-1}(\delta(x))]^{-d}u(x)\,\} +$$

$$+ \sup_{x\in\Gamma_{\varphi,\rho_0}(\zeta,\alpha)} \{\,\delta(x)^{N+\gamma}[\varphi^{-1}(\delta(x))]^{-d}u(x)\,\},$$

we obtain:

$$\int_{\partial G} \sup_{x\in\Gamma_{\varphi}(\zeta,\alpha)} \{\delta(x)^{N+\gamma}[\varphi^{-1}(\delta(x))]^{-d}u(x)\}\, d\mathcal{H}^d(\zeta) \leq$$

$$\leq \int_{\partial G} \sup_{x\in\Gamma^c_{\varphi,\rho_0}(\zeta,\alpha)} \{\delta(x)^{N+\gamma}[\varphi^{-1}(\delta(x))]^{-d}u(x)\}\, d\mathcal{H}^d(\zeta)+$$

$$+\int_{\partial G} \sup_{x\in\Gamma_{\varphi,\rho_0}(\zeta,\alpha)} \{\delta(x)^{N+\gamma}[\varphi^{-1}(\delta(x))]^{-d}u(x)\}\, d\mathcal{H}^d(\zeta) \leq$$

$$\leq \int_{\partial G} \sup_{x\in\Gamma^c_{\varphi,\rho_0}(\zeta,\alpha)} \{\delta(x)^{N+\gamma}[\varphi^{-1}(\delta(x))]^{-d}u(x)\}\, d\mathcal{H}^d(\zeta)+C\int_{G_{\rho'_0}} \delta(x)^{\gamma}u(x)\, dm_N(x) \leq$$

$$\leq \int_{\partial G} \sup_{x\in\Gamma^c_{\varphi,\rho_0}(\zeta,\alpha)} \{\delta(x)^{N+\gamma}[\varphi^{-1}(\delta(x))]^{-d}u(x)\}\, d\mathcal{H}^d(\zeta)+C\int_{G} \delta(x)^{\gamma}u(x)\, dm_N(x).$$

It remains to show that

$$\int_{\partial G} \sup_{x\in\Gamma^c_{\varphi,\rho_0}(\zeta,\alpha)} \{\delta(x)^{N+\gamma}[\varphi^{-1}(\delta(x))]^{-d}u(x)\}\, d\mathcal{H}^d(\zeta) \leq C\int_{G} \delta(x)^{\gamma}u(x)\, dm_N(x)$$

for some $C = C(N,G,d,\varphi,\alpha,\gamma,K')$. For all $x \in \Gamma^c_{\varphi,\rho_0}(\zeta,\alpha)$ we have

$$u(x) \leq \frac{K'}{\nu_N(\frac{\delta(x)}{3})^N}\int_{B(x)} u(y)\, dm_N(y).$$

Using also the facts that $\frac{2}{3}\delta(x) \leq \delta(y) \leq \frac{4}{3}\delta(x)$ for all $y \in B(x)$, one gets easily:

$$\int_{\partial G} \sup_{x\in\Gamma^c_{\varphi,\rho_0}(\zeta,\alpha)} \{\delta(x)^{N+\gamma}[\varphi^{-1}(\delta(x))]^{-d}u(x)\}\, d\mathcal{H}^d(\zeta) \leq$$

$$\leq \int_{\partial G} \sup_{x\in\Gamma^c_{\varphi,\rho_0}(\zeta,\alpha)} \{\delta(x)^{N+\gamma}[\varphi^{-1}(\delta(x))]^{-d}\frac{K'}{\nu_N(\frac{\delta(x)}{3})^N}\int_{B(x)} u(y)\, dm_N(y)\}d\mathcal{H}^d(\zeta) \leq$$

$$\leq \frac{3^N K'}{\nu_N}\int_{\partial G} \sup_{x\in\Gamma^c_{\varphi,\rho_0}(\zeta,\alpha)} \{\delta(x)^{\gamma}[\varphi^{-1}(\delta(x))]^{-d}\int_{B(x)} u(y)\, dm_N(y)\}d\mathcal{H}^d(\zeta) \leq$$

$$\leq \left(\frac{3}{2}\right)^{|\gamma|}\frac{3^N K'}{\nu_N}\int_{\partial G} \sup_{x\in\Gamma^c_{\varphi,\rho_0}(\zeta,\alpha)} \{[\varphi^{-1}(\delta(x))]^{-d}\int_{B(x)} \delta(y)^{\gamma}u(y)\, dm_N(y)\}d\mathcal{H}^d(\zeta) \leq$$

$$\leq \frac{3^{|\gamma|+N}K'}{2^{|\gamma|}\nu_N}[\varphi^{-1}(\rho_0)]^{-d}\mathcal{H}^d(\partial G)\int_G \delta(y)^{\gamma}u(y)\, dm_N(y).$$

Thus

$$\int_{\partial G} \sup_{x\in\Gamma_\varphi(\zeta,\alpha)} \{\delta(x)^{N+\gamma}[\varphi^{-1}(\delta(x))]^{-d}u(x)\}\,d\mathcal{H}^d(\zeta) \leq C\int_G \delta(x)^\gamma u(x)\,dm_N(x),$$

concluding the proof. □

Corollary 7.14. *Let* $u : B^N \to [0,+\infty)$ *be a subharmonic function. Let* $p > 0$, $\alpha > 1$, *and* $\gamma > -1-\beta(p)$, *where* $\beta(p) := \max\{(N-1)(1-p),0\}$. *Then there is a constant* $C = C(N,p,\alpha,\gamma)$ *such that*

$$\int_{S^{N-1}} \sup_{x\in\Gamma_{id}(\zeta,\alpha)} \{(1-|x|)^{\gamma+1}u(x)^p\}\,d\sigma(\zeta) \leq C\int_{B^N}(1-|x|)^\gamma u(x)^p\,dm_N(x).$$

Here id is the identity mapping of $\mathbb{R}^N$ *and* σ *is the spherical (Lebesgue) measure in* S^{N-1}.

Remark 7.15. For related results in locally uniformly spaces, see [114], Theorem 7.5 and Corollary 7.6, pp. 28–36. See also [86].

7.4. **A limiting case result of a nonintegrability result of Suzuki.**

7.4.1. ***Suzuki's result.*** Suzuki [146], Theorem and its proof, pp. 113–115, gave the following result.

Theorem 7.16. *Let* $0 < p \leq 1$. *If a superharmonic (resp. nonnegative subharmonic) function* v *on* D *satisfies*

$$\int_D |v(x)|^p\,\delta(x)^{Np-N-2p}\,dm_N(x) < +\infty \tag{7.8}$$

then v *vanishes identically.*

7.4.2. ***Our improvement.*** Suzuki pointed also out that his result is sharp in the following sense: If p, $0 < p \leq 1$, is fixed, then the exponent $\gamma = Np-N-2p$ in (7.8) cannot be increased. On the other hand, clearly $-N < \gamma \leq -2$, when $0 < p \leq 1$. Since the class of permissible functions include, in addition the functions t^p, $0 < p \leq 1$, also a large amount of essentially different functions (see 1.3 above), one is tempted to ask whether there exists any limiting ("borderline") case result for Suzuki's results, corresponding to the case $p = 0$. To be more precise, and replacing Suzuki's condition (7.8) above with a slightly more general condition, we pose the following question:

Let D *and* v *be as above. Let* $\gamma \leq -N$ *and let* $\psi : [0,+\infty) \to [0,+\infty)$ *be a permissible function. Does the condition*

$$\int_D \psi(|v(x)|)\delta(x)^\gamma\,dm_N(x) < +\infty,$$

imply $v \equiv 0$?

Observe that the least severe form of above integrability condition occurs when $\gamma = -N$.

Below in Corollary 7.19 we answer the question in the affirmative, in the case of any strictly permissible function ψ. In order to achieve this, we first formulate below in Theorem 7.18 a general result for arbitrary $\gamma \leq -2$ which is, for $-N < \gamma \leq -2$, however, essentially more or less just Suzuki's above result (see Remark 7.20 (b) below). Our formulation has the advantage that, unlike Suzuki's result, it contains a certain limiting case, Corollary 7.19, too.

The proof which we below write down (and in quite detail, just for the convenience of the reader) is merely a slight modification of Suzuki's argument, combined with our version for the generalized mean value inequality (Proposition 1.3 (iv) above), and also some additional estimates.

Lemma 7.17. *Let u be a nonnegative subharmonic function on D. Suppose* $\psi : [0,+\infty) \to [0,+\infty)$ *is a permissible function such that*

$$\int_D \psi(u(x))\delta(x)^\gamma dm_N(x) < +\infty$$

for some $\gamma \in \mathbb{R}$. *Then* $\psi(u(x)) = o(\delta(x)^{-N-\gamma})$ *as* $\delta(x) \to 0$.

Proof. By Proposition 1.3 (iv), $\psi \circ u$ is quasinearly subharmonic on D. Write for $x \in D$, $B_0 = B^N(x, \frac{1}{2}\delta(x))$. Since

$$\frac{1}{2}\delta(x) < \delta(y) < \frac{3}{2}\delta(x), \tag{7.9}$$

for all $y \in B_0$, we get

$$\begin{aligned}\psi(u(x)) &\leq 2^{N+|\gamma|} C_0\,\delta(x)^{-N-\gamma} \int_{B_0} \psi(u(y))\delta(y)^\gamma dm_N(y) \leq \\ &\leq C\,\delta(x)^{-N-\gamma} \int_{D_{\delta(x)}} \psi(u(y))\delta(y)^\gamma dm_N(y),\end{aligned}$$

where $C = C(\gamma, N, \psi, u) > 0$. The claim follows. □

Theorem 7.18. ([101], Theorem 3, p. 178, and [100], Theorem 4.3, p. 7) *Let D be bounded. Let v be a superharmonic (resp. nonnegative subharmonic) function on D. Let* $\psi : [0,+\infty) \to [0,+\infty)$ *be a strictly permissible function. Suppose*

$$\int_D \psi(|v(x)|)\delta(x)^\gamma dm_N(x) < +\infty, \tag{7.10}$$

where $\gamma \leq -2$ *is such that there is a constant* $C = C(\gamma, N, \psi, D) > 0$ *such that*

$$s^{N+\gamma} \leq \psi(C s^{N-2}) \text{ for all } s > \frac{1}{diam D}. \tag{7.11}$$

Then v vanishes identically.

Proof. We write the proof down only for the case $N \geq 3$. Write $v^+ = \max\{v, 0\}$ and $s = v^- = -\min\{v, 0\}$. Then $|v| = v^+ + v^-$ and $s \geq 0$ is subharmonic. (Resp. if v is nonnegative and subharmonic, let $s = v$.) Proceeding as Suzuki, but using also some additional estimates, we will show that $s \equiv 0$.

By (7.10),

$$\int_D \psi(s(x))\delta(x)^\gamma dm_N(x) < +\infty,$$

thus

$$v_{\psi,\gamma}(x) = \int_D G_D(x,y)\psi(s(y))\delta(y)^\gamma dm_N(y) < +\infty$$

is a potential by [48], Theorem 6.3, p. 99. Here G_D is the Green function of D.

By Lemma 7.17, $\psi(s(y)) \leq C\delta(y)^{-N-\gamma}$. Thus also

$$s(y) \leq \psi^{-1}(C\delta(y)^{-N-\gamma}) \leq C\psi^{-1}(\delta(y)^{-N-\gamma}) \tag{7.12}$$

for all $y \in D$, where $C = C(\gamma, N, \psi, s, D) \geq 1$. Let $x \in D$ be fixed for a while. Let $B = B^N(x, \delta(x))$ and $B_0 = B^N(x, \frac{1}{2}\delta(x))$. Using (7.12) (and (7.9)) one gets

$$s(y) \leq C\psi^{-1}(\delta(y)^{-N-\gamma}) \leq C\psi^{-1}(2^{N+|\gamma|}\delta(x)^{-N-\gamma}) \leq C\psi^{-1}(\delta(x)^{-N-\gamma})$$

for all $y \in B_0$. Therefore

$$\frac{\psi(s(y))}{s(y)} \geq C_1 \frac{\psi(C\psi^{-1}(\delta(x)^{-N-\gamma}))}{C\psi^{-1}(\delta(x)^{-N-\gamma})} \geq C\frac{\delta(x)^{-N-\gamma}}{\psi^{-1}(\delta(x)^{-N-\gamma})}$$

for all $y \in B_0$. With the aid of this and of a standard estimate for the Green function $G_B(x,\cdot)$ in B_0 and of (7.12), one gets

$$\begin{aligned} v_{\psi,\gamma}(x) &= \int_{D\Omega} G_D(x,y)\psi(s(y))\delta(y)^\gamma dm_N(y) \geq \int_{B_0} G_B(x,y)\psi(s(y))\delta(y)^\gamma dm_N(y) \geq \\ &\geq \int_{B_0} G_B(x,y)s(y)\frac{\psi(s(y))}{s(y)}\delta(y)^\gamma dm_N(y) \geq \\ &\geq C\frac{\delta(x)^{-N-\gamma}}{\psi^{-1}(\delta(x)^{-N-\gamma})}\int_{B_0} |x-y|^{2-N}\delta(y)^\gamma s(y)\, dm_N(y) \geq \\ &\geq C\frac{1}{\delta(x)^{N-2}\psi^{-1}(\delta(x)^{-N-\gamma})}s(x). \end{aligned}$$

By (7.11) we see that there is a constant $C_3 \geq 1$ such that

$$\delta(y)^{N-2}\psi^{-1}(\delta(y)^{-N-\gamma}) \leq C_3$$

for all $y \in D$. Combining this with the above estimate for $v_{\psi,\gamma}$, one gets

$$v_{\psi,\gamma}(x) \geq C s(x),$$

where $C = C(\gamma, N, \psi, s, D) > 0$. Remembering that $x \in D$ was arbitrary, that $v_{\psi,\gamma}$ is a potential and s subharmonic, it follows from [48], Corollary 6.19, p. 117, that $s \equiv 0$. Thus $v = v^+ \geq 0$. It remains to show that $v \equiv 0$.

As above,

$$\begin{aligned} \int_D G_D(x,y)\,\delta(y)^{-2} dm_N(y) &\geq \int_{B_0} G_B(x,y)\,\delta(y)^{-2} dm_N(y) \geq \\ &\geq C\int_{B_0} |x-y|^{2-N}\,\delta(y)^{-2} dm_N(y) \geq \\ &\geq C\delta(x)^{2-N}\,\delta(x)^{-2}\nu_N\,(\frac{\delta(x)}{2})^N = C, \end{aligned}$$

where $C = C(N) > 0$. Thus by [48], Lemma 6.1, p. 98, and Corollary 6.19, p. 117,

$$\int_D G_D(x,y)\,\delta(x)^{-2}\,dm_N(y) = +\infty \tag{7.13}$$

for all $x \in D$. Consider next an arbitrary potential w on D,

$$w(x) = \int_D G_D(x,y)\,d\lambda(y) \tag{7.14}$$

where $\lambda \neq 0$ is a measure on D. From (7.13) it follows that

$$\begin{aligned}\int_D w(x)\,\delta(x)^{-2}\,dm_N(x) &= \int_D \Big[\int_D G_D(x,y)\,d\lambda(y)\Big]\,\delta(x)^{-2}\,dm_N(x) = \\ &= \int_D \Big[\int_D G_D(x,y)\,\delta(x)^{-2} dm_N(x)\Big]\,d\lambda(y) = +\infty.\end{aligned}$$

Using this and the facts that D is bounded and w, as a superharmonic function, is locally integrable, one sees that

$$\int_{D_1} w(x)\,\delta(x)^{-2}\,dm_N(x) = +\infty \tag{7.15}$$

where $D_1 = \{x \in D : \delta(x) < 1\}$.

Suppose in particular that w in (7.14) is the potential of the superharmonic function $v_M = \inf\{v, M\}$, where $M \geq 0$. Then $v_M \geq w$, and one has by (7.15) and by the fact that $\gamma \leq -2$,

$$\begin{aligned}+\infty > \int_D \psi(v(x))\,\delta(x)^{\gamma} dm_N(x) &\geq \int_D \psi(v_M(x))\,\delta(x)^{\gamma} dm_N(x) \geq \\ &\geq \int_D v_M(x)\frac{\psi(v_M(x))}{v_M(x)}\,\delta(x)^{\gamma} dm_N(x) \geq \\ &\geq C\frac{\psi(M)}{M}\int_D v_M(x)\,\delta(x)^{\gamma} dm_N(x) \geq \\ &\geq C\frac{\psi(M)}{M}\int_{D_1} w(x)\,\delta(x)^{-2}\,dm_N(x) = +\infty,\end{aligned}$$

a contradiction unless $w \equiv 0$. Since $w \equiv 0$, the nonnegative superharmonic functions v_M, $M \geq 0$, are in fact harmonic, e.g. by the Riesz Decomposition Theorem [48], Theorem 6.18, p. 116. Since this is impossible, one has $v \equiv 0$, concluding the proof. □

Corollary 7.19. *Let D be bounded. Let v be a superharmonic (resp. nonnegative subharmonic) function on D. Let $\psi : [0, +\infty) \to [0, +\infty)$ be any strictly permissible function and let $\gamma \leq -N$. If*

$$\int_D \psi(|v(x)|)\delta(x)^{\gamma} dm_N(x) < +\infty,$$

then v vanishes identically.

Proof. For the proof observe that the condition (7.11) is indeed satisfied for $\gamma \leq -N$, since D is bounded and ψ is increasing. □

Remark 7.20. Next we consider the assumptions made in Theorem 7.18.

(a) Our assumption $\gamma \leq -2$ is unnecessary, and it could be dropped: If $\gamma \in \mathbb{R}$, then it follows easily from (7.11) and from the property (c) in 1.3 of strictly permissible functions that indeed $\gamma \leq -2$.

(b) Suppose that $-N < \gamma \leq -2$. If, instead of (7.11), one supposes that

$$s^{N+\gamma} \leq \psi(C s^{N-2}) \text{ for all } s > 0,$$

then clearly

$$\psi(|v(x)|) \geq C^{-\frac{N+\gamma}{N-2}} |v(x)|^{\frac{N+\gamma}{N-2}}$$

for all $x \in D$. Thus (7.10) implies that

$$\int_D |v(x)|^{\frac{N+\gamma}{N-2}} \delta(x)^\gamma dm_N(x) < +\infty,$$

and hence $v \equiv 0$ by Suzuki's result, Theorem 7.16 above. Recall that here $0 < p = \frac{N+\gamma}{N-2} \leq 1$ and $\gamma = Np - N - 2p$. Thus Theorem 7.18, but now the assumption (7.11) replaced with the aforesaid assumption, is just a restatement of Suzuki's result for bounded domains.

(c) If $\gamma \leq -N$, then the condition (7.11) clearly holds, since ψ is strictly permissible. This case gives indeed the already referred limiting case for Suzuki's result. See Corollary 7.19 above.

Remark 7.21. The result of Theorem 7.18 does not, of course, hold any more, if one replaces strictly permissible functions by permissible functions. For a counter example, set, say, $v(x) = |x|^{2-N}$, $\psi(t) = t^p$, where $\frac{N-1}{N-2} < p < \frac{N}{N-2}$, $\gamma = Np - N - 2p$ or just $\gamma > 1$. Then clearly

$$\int_B v(x)^p \delta(x)^\gamma dm_N(x) < +\infty$$

but $v \not\equiv 0$.

Remark 7.22. Provided D is bounded and ψ is strictly permissible, one can, with the aid of Theorem 7.18 and Corollary 7.19, exclude some trivial cases $u \equiv 0$ from the result of Theorem 7.5 by imposing certain restrictions on the exponent γ. We point out only two cases:

(i) By Corollary 7.19, $\gamma > -N$, regardless of ψ.

(ii) By Suzuki's result, Theorem 7.16 above, $\gamma > Np - N - 2p$, in the case when $\psi(t) = t^p$, $0 < p \leq 1$.

8. Minkowski content and removable sets for subharmonic functions

Abstract. It is an open problem whether a function, subharmonic with respect to the first variable and harmonic with respect to the second, is subharmonic or not. Based again on our mean value type inequality, we improve our previous subharmonicity results of the above type functions, thus improving also the previous results of Kołodziej and Thorbiörnson and Imomkulov. Moreover, we give refinements, with concise proofs, to the older basic results of Arsove, and of Cegrell and Sadullaev.

Keywords. Subharmonic, quasinearly subharmonic, separately quasinearly subharmonic, harmonic, net measure, Minkowski content

8.1. **Previous results.** It is a classical result [46], Theorem 5.18, p. 237, that if u is subharmonic in $D\setminus E$ and bounded above and moreover E is polar, then u has a subharmonic extension to the whole of D. Imposing certain constraints on the geometry and size of the set E, Gardiner relaxed considerably the boundedness requirement of u, see [37], Theorem 1 and Theorem 3, pp. 71–74 (for these results, see also [98], Theorem A and Theorem B, p. 199). To state his results, let $\Phi : D \to \mathbb{R}$ be a $\mathcal{C}^2$ function with nonvanishing gradient throughout D. Put $S := \{x \in D : \Phi(x) = 0\}$. Write $d(x,S)$ for the distance from $x \in \mathbb{R}^N$ to S and let $\mathcal{H}^\alpha$ be the α-dimensional Hausdorff outer measure in $\mathbb{R}^N$.

Theorem 8.1. *Let $\alpha \in (0,N-2)$ and E be a compact subset of S such that $\mathcal{H}^\alpha(E)=0$. If u is subharmonic in $D\setminus E$ and satisfies*

$$u(x) \leq C d(x,S)^{\alpha+2-N} \quad (x \in D\setminus S)$$

for some positive constant C, then u has a subharmonic extension to D.

Theorem 8.2. *Let $\alpha \in (0,N-2)$ and E be a compact subset of S such that $\mathcal{H}^\alpha(E) < +\infty$. If u is subharmonic in $D\setminus E$ and satisfies*

$$u(x) \leq v(d(x,S)) \quad (x \in D\setminus S),$$

where $t^{N-2-\alpha}v(t) \to 0+$ $(t \to 0+)$, then u has a subharmonic extension to D.

Gardiner also shows [37], Theorem 2 and Theorem 4, pp. 72–73, that his results are sharp in the following sense: If one drops the smoothness assumption $E \subset S$ then the exceptional set E is not any more necessarily removable. Our purpose is to point out that there exist, however, results which are in a certain sense parallel to Gardiner's results but where no smoothness conditions are necessary to impose on the exceptional set. As a matter of fact, we show below in Theorem 8.5 and Theorem 8.6 that results similar to Gardiner's hold when his conditions

(i) *$E \subset S$ where S is a $\mathcal{C}^2$ $(N-1)$-dimensional manifold in D,*
(ii) *$\mathcal{H}^\alpha(E) = 0$ (resp. $\mathcal{H}^\alpha(E) < +\infty$),*

Juhani Riihentaus

are replaced by one geometric Minkowski measure condition $\mathcal{M}^{\alpha}(E) = 0$ (resp. $\mathcal{M}^{\alpha}(E) < +\infty$). Our proofs are different and perhaps shorter than those of Gardiner. Moreover, our approach does not require the exceptional set E to be compact, unlike Gardiner's results. On the other hand, as is shown in Example 16 and Example 17 below, Gardiner's and our results are independent: Neither our nor Gardiner's results are included in the other's.

Gardiner also [37], Theorem 5, p. 74 (see also [98], Theorem C, p. 200), proves the following result:

Theorem 8.3. *Let* $\alpha \in (0, N-2)$ *and* E *be a compact subset of* S *such that* $\mathcal{H}^{\alpha}(E) = 0$. *If* u *is subharmonic in* $D \setminus E$ *and satisfies*

$$\mathcal{A}(u^+, x, r) \le C\, r^{\alpha+2-N} \quad (\overline{B^N(x,r)} \subset D)$$

for some positive constant C, *then* u *has a subharmonic extension to* D.

Below in Theorem 8.7 we improve this result by dropping the condition that E is compact. Again our approach is essentially different than that of Gardiner.

8.2. **Net measure and Minkowski content.** For readers' convenience we first recall certain basic facts concerning net measure and Minkowski content and their relationship with the standard Hausdorff measure. For a more thorough discussion see e.g. [44], pp. 41–44, and [34], pp. 33, 42.

Let $A \subset \mathbb{R}^N$ and $\alpha \in [0, N]$. For each $\varepsilon > 0$ define

$$\mathcal{L}_{\alpha}^{\varepsilon}(A) = \inf \sum_{i=1}^{+\infty} s_i^{\alpha}$$

where the infimum is over all coverings of A by countable disjoint collection of dyadic cubes Q_i with (side)length $s_i \le \varepsilon$. Define the *α-dimensional net measure* of A by

$$\mathcal{L}_{\alpha}(A) = \lim_{\varepsilon \to 0+} \mathcal{L}_{\alpha}^{\varepsilon}(A).$$

It is well-known that the standard Hausdorff measure $\mathcal{H}^{\alpha}$ and the net measure $\mathcal{L}_{\alpha}$ are comparable: There are positive constants $C_1 = C_1(N)$ and $C_2 = C_2(N)$ such that

$$C_1 \mathcal{L}_{\alpha}(A) \le \mathcal{H}^{\alpha}(A) \le C_2 \mathcal{L}_{\alpha}(A) \tag{8.1}$$

for all $A \subset \mathbb{R}^N$.

To define the Minkowski content, let $A \subset \mathbb{R}^N$, $\alpha \in [0, N]$ and $\varepsilon > 0$. Write

$$A_{\varepsilon} = \{x \in \mathbb{R}^N : d(x, A) < \varepsilon\}.$$

The *α-dimensional upper Minkowski content* of A is defined by

$$\mathcal{M}^{\alpha}(A) = \limsup_{\varepsilon \to 0+} \frac{m_N(A_{\varepsilon})}{\varepsilon^{N-\alpha}}.$$

It is well-known that there is a positive constant $C_3 = C_3(N, \alpha)$ such that

$$C_3 \mathcal{H}^{\alpha}(A) \le \mathcal{M}^{\alpha}(A)$$

for all $A \subset \mathbb{R}^N$. The reverse inequality does not hold in general, but is true for certain smooth sets, even for α rectifiable closed subsets of $\mathbb{R}^N$ (here α is a positive integer). See [44], p. 41, and [35], 3.2.39, p. 275.

Our argument will essentially be based on the following type of partition of unity, see [44], Lemma 3.1, p. 43:

Lemma 8.4. *Let $\{Q_i : i = 1, 2, \ldots, N^*\}$ be a finite disjoint collection of dyadic cubes of length $s(Q_i) = s_i$. For each i, there exists a function $\varphi_i \in \mathcal{C}_0^\infty(\mathbb{R}^N)$ with support* $\operatorname{spt}\varphi_i \subset \frac{3}{2}Q_i$ *such that $\sum_{i=1}^{+\infty} \varphi_i(x) = 1$ for all $x \in \cup_{i=1}^{N^*} Q_i$. Furthermore, for each multi-index λ, there is a constant $C_\lambda = C_\lambda(\lambda, N)$ for which $|\mathcal{D}^\lambda \varphi_i(x)| \leq C_\lambda s_i^{-|\lambda|}$ for all $x \in \mathbb{R}^N$ and $i = 1, 2, \ldots, N^*$.*

8.3. **The results.** Our first result is parallel to Gardiner's cited result, Theorem 8.1 above:

Theorem 8.5. ([98], Theorem 1, p. 201) *Suppose that $\alpha \in [0, N-2]$ and $\mathcal{M}^\alpha(E) = 0$. If f is subharmonic in $\Omega \setminus E$ and satisfies*

$$f(x) \leq C^* \, d(x, E)^{\alpha+2-N} \quad (x \in \Omega \setminus E)$$

for some positive constant C^*, then f has a subharmonic extension to Ω.

Proof. If $\alpha = 0$ then $E = \emptyset$. If $\alpha = N-2$, then E is polar e.g. by [46], Theorem 5.14, p. 288. Since f is then also bounded above, the claim follows from the classical result [46], Theorem 5.18, p. 237.

It remains to consider the case $\alpha \in (0, N-2)$. Since f^+ is subharmonic, and also

$$f^+(x) \leq C^* d(x, E)^{\alpha+2-N} \qquad (x \in \Omega \setminus E),$$

we may suppose that $f \geq 0$.

We first show that $f \in \mathcal{L}^1_{\mathrm{loc}}(\Omega)$, cf. [44], p. 42, and [96], pp. 730-731. It is sufficient to show that for some $r > 0$,

$$\int_{E_r} f \, dm_N < +\infty.$$

Take $\varepsilon > 0$ arbitrarily. Since $\mathcal{M}^\alpha(E) = 0$, there is r_0, $0 < r_0 < 1$, such that $m_N(E_r) \leq \varepsilon r^{N-\alpha}$ for all r, $0 < r \leq r_0$. Take any such r, and write for each $j = 0, 1, \ldots,$

$$K_j = \{x \in \mathbb{R}^N : d(x, E) < r 2^{-j}\}.$$

Then

$$E_r = \cup_{j=0}^{+\infty} (K_j \setminus K_{j+1}),$$

and

$$\begin{aligned}\int_{E_r} f(x)\,dm_N(x) \le & C^* \int_{E_r} d(x,E)^{\alpha+2-N}\,dm_N(x) = \\ = & C^* \sum_{j=0}^{+\infty} \int_{K_j \setminus K_{j+1}} d(x,E)^{\alpha+2-N}\,dm_N(x) \le \\ \le & C^* \sum_{j=0}^{+\infty} \left[r2^{-(j+1)}\right]^{\alpha+2-N} m_N(K_j) \le \\ \le & C^* 2^{N-2-\alpha} r^{\alpha+2-N} \sum_{j=0}^{+\infty} 2^{(N-2-\alpha)j} \varepsilon (r2^{-j})^{N-\alpha} \le \\ \le & C^* 2^{N-2-\alpha} r^2 \varepsilon \sum_{j=0}^{+\infty} 2^{-2j} < +\infty.\end{aligned}$$

Thus $f \in \mathcal{L}^1_{\mathrm{loc}}(\Omega)$. For later use we observe that we also got

$$\int_{E_r} d(x,E)^{\alpha+2-N}\,dm_N(x) \le C r^2 \varepsilon \tag{8.2}$$

for all r, $0 \le r \le r_0$, where $C = C(N,\alpha,C^*)$.

To complete the proof, it remains to show that for any nonnegative test function $\varphi \in \mathcal{D}(\Omega)$,

$$\int f \Delta\varphi\, dm_N \ge 0.$$

We may suppose that $0 \le \varphi \le 1$ and $|\mathcal{D}^\lambda \varphi| \le 1$ for each multi-index λ, $|\lambda| \le 2$. Compare [57], p. 113.

Let $K = \operatorname{spt}\varphi$. We may suppose that $K_{r_0} \subset \Omega$. Choose $s = 2^{-k}$ so small that $3s\sqrt{N} \le r_0$. Cover K by a finite, disjoint collection of dyadic cubes Q_i with length $s(Q_i) = s$, $i = 1,2,\ldots,N_1$. We may suppose that

$$\frac{3}{2}Q_i \cap E \neq \emptyset \text{ for } i = 1,2,\ldots,N^*, \tag{8.3}$$

and

$$\frac{3}{2}Q_i \cap E = \emptyset \text{ for } i = N^*+1, N^*+2,\ldots,N_1, \tag{8.4}$$

for some $N^* \in \mathbb{N}$, $1 \le N^* \le N_1$. Let φ_i, $i = 1,2,\ldots,N_1$, be the test functions related to the collection Q_i, $i = 1,2,\ldots,N_1$, and possessing the properties described in the above Lemma. Since f is subharmonic in $\Omega \setminus E$ and all $\varphi\varphi_i$, $i = N^*+1, N^*+2,\ldots,N_1$, are nonnegative test functions in $\mathcal{D}(\Omega \setminus E)$, we have

$$\int f \Delta(\varphi\varphi_i)\,dm_N \ge 0 \text{ for } i = N^*+1, N^*+2,\ldots,N_1.$$

In view of these inequalities we get

$$
\begin{aligned}
\int f\Delta\varphi\, dm_N = \int f\Delta\left[\varphi\left(\sum_{j=1}^{N_1}\varphi_j\right)\right] dm_N = \sum_{i=1}^{N_1}\int_{\frac{3}{2}Q_i} f\,\Delta(\varphi\varphi_j)\, dm_N \geq \\
\geq \sum_{i=1}^{N^*}\int_{\frac{3}{2}Q_i} f\,\Delta(\varphi\varphi_j)\, dm_N.
\end{aligned}
\tag{8.5}
$$

An easy computation shows that

$$
\Delta(\varphi\varphi_i) = (\Delta\varphi)\varphi_i + \varphi(\Delta\varphi_i) + 2\sum_{j=1}^{N_1}\mathcal{D}_j\varphi\mathcal{D}_j\varphi_i.
$$

By the properties of the test functions φ_i and φ, we have for all $i = 1,2,\ldots,N^*$ and $x \in \mathbb{R}^N$,

$$
\begin{aligned}
|\Delta(\varphi\varphi_i)(x)| \leq & |\Delta\varphi(x)||\varphi_i(x)| + |\varphi(x)||\Delta\varphi_i(x)| + 2\sum_{j=1}^{N_1}|\mathcal{D}_j\varphi(x)||\mathcal{D}_j\varphi_i(x)| \leq \\
\leq & 1 + \frac{C_2}{s^2} + \frac{C_1}{s} \leq \frac{C}{s^2},
\end{aligned}
\tag{8.6}
$$

where $C = C(N, C_1, C_2)$. The last inequality here follows from the fact that, since $0 < r_0 < 1$, also $0 < s < 1$.

For each cube Q_i, $i = 1,2,\ldots,N^*$, there are clearly at most 3^N cubes Q_j, $s(Q_j) = s$, $j = 1,2,\ldots,N_i \leq 3^N$ (just the adjacent cubes to Q_i with equal length) such that

$$
\frac{3}{2}Q_i \cap \frac{3}{2}Q_j \neq \emptyset. \tag{8.7}
$$

Using this, the fact that $\frac{3}{2}Q_i \subset E_{3s\sqrt{N}}$, $i = 1,2,\ldots,N^*$, (8.5) and (8.6), we get

$$
\begin{aligned}
\int f\Delta\varphi\, dm_N \geq -\frac{C}{s^2}\sum_{i=1}^{N^*}\int_{\frac{3}{2}Q_i} f\, dm_N = -\frac{C}{s^2}\sum_{i=1}^{N^*}\int_{E_{3s\sqrt{N}}} f\chi_{\frac{3}{2}Q_i} dm_N = \\
\geq -\frac{C}{s^2}\int_{E_{3s\sqrt{N}}} f\left(\sum_{i=1}^{N^*}\chi_{\frac{3}{2}Q_i}\right) dm_N \geq -\frac{3^N C}{s^2}\int_{E_{3s\sqrt{N}}} f\, dm_N.
\end{aligned}
$$

Here $\chi_{\frac{3}{2}Q_i}$ is the characteristic function of $\frac{3}{2}Q_i$, $i = 1,2,\ldots,N^*$. Above we have used the fact that $\sum_{i=1}^{N^*}\chi_{\frac{3}{2}Q_i}(x) \leq 3^N$ for all $x \in E_{3s\sqrt{N}}$. Indeed, if $x \notin \frac{3}{2}Q_i$ for $i = 1,2,\ldots,N^*$, then $\sum_{i=1}^{N^*}\chi_{\frac{3}{2}Q_i}(x) = 0$. If $x \in \frac{3}{2}Q_{i_0}$ for some i_0, $1 \leq i_0 \leq N^*$, then by (8.7) we see that among the cubes $\frac{3}{2}Q_i$, $i = 1,2,\ldots,N^*$, there are at most N_{i_0} such for which $x \in \frac{3}{2}Q_i$. Since $N_{i_0} \leq 3^N$ (see (8.4) above), also $\sum_{i=1}^{N^*}\chi_{\frac{3}{2}Q_i}(x) \leq 3^N$. Proceeding then further, and using also (8.2), we get

$$\begin{aligned}\int f\Delta dm_N &\geq -\frac{C}{s^2}\int_{E_{3s\sqrt{N}}} f\,dm_N \geq \\ &\geq -\frac{C}{s^2}\int_{E_{3s\sqrt{N}}} d(x,E)^{\alpha+2-N}\,dm_N(x) \geq \\ &\geq -\frac{C}{s^2}(3s\sqrt{N})^2\varepsilon = -C\varepsilon.\end{aligned}$$

Since $\varepsilon > 0$ was arbitrary and $C = C(N,\alpha,C^*)$, it follows that

$$\int f\Delta\varphi\,dm_N \geq 0,$$

concluding the proof. □

As Gardiner points out [37], p. 73, a slight modification of his proof of Theorem A yields Theorem B. In our frame the situation is similar.

Theorem 8.6. ([98], Theorem 2, p. 205) *Let* $\alpha \in [0,N-2]$ *and* $\mathcal{M}^\alpha(E) < +\infty$. *If* f *is subharmonic in* $\Omega\setminus E$ *and satisfies*

$$f(x) \leq u(d(x,E)) \quad (x \in \Omega\setminus E),$$

where $u(t)$ *is a Borel measurable function such that* $t^{N-2-\alpha}u(t) \to 0$ $(t \to 0+)$, *then* f *has a subharmonic extension to* Ω.

Proof. The proof goes along the same lines as above with only minor changes. In fact, take $\varepsilon > 0$ arbitrarily. Choose then r_0, $0 < r_0 < 1$, such that

$$u(t) < \varepsilon t^{\alpha+2-N}$$

whenever $0 < t < r_0$. Since $\mathcal{M}^\alpha(E) < +\infty$, we may suppose that $m_N(E_r) < Mr^{N-\alpha}$ for all r, $0 < r \leq r_0$. Proceeding then as in the proof of Theorem 8.5 (see (8.1) above), one sees that for all r, $0 < r < r_0$,

$$\begin{aligned}\int_{E_r} f(x)\,dm_N(x) \leq \int_{E_r} u(d(x,E))\,dm_N(x) \leq \int_{E_r} d(x,E)^{\alpha+2-N}\,dm_N(x) \leq \\ \leq \varepsilon C r^2 M < +\infty.\end{aligned}$$

The rest of the proof goes as in the proof of Theorem 8.5. □

Example 16. Let $0 < \alpha < 1$ be arbitrarily given. By [34], Example 4.5, p. 58, there is a uniform Cantor set $F \subset [0,1]$ such that $\mathcal{M}^\alpha(F) = 0$. Set $E = F \times F \times \cdots \times F$. Then E is closed and by [34], Example 7.6, p. 95, $\mathcal{M}^\alpha(E) = 0$. Clearly E is not contained in any $\mathcal{C}^2$ $(N-1)$-dimensional manifold. Thus our results, Theorems 8.5 and 8.6 above, can be applied in situations where Gardiner's above results, Theorem 8.1 and Theorem 8.2 cannot be used.

Example 17. By [62], 2.3, p. 462, there is for each α, $0 < \alpha < 2$, a countable, compact subset F of the complex plane $\mathbb{C}$ with $\mathcal{M}^{\alpha}(F) > 0$. Let $E = F \times \{0\} \subset \mathbb{R}^3$. One sees easily that $\mathcal{M}^{\alpha}(E) > 0$. Since E is countable, $\Lambda_{\lambda}(E) = 0$. Thus we have an example where Gardiner's theorems can be used whereas our results are not applicable.

Our last theorem improves Gardiner's result [37], Theorem 5, p. 74 (see also [98], Theorem C, p. 200) by allowing the exceptional set to be noncompact. The proof we present is different from that of Gardiner.

Theorem 8.7. ([98], Theorem 3, p. 205) *Let* $\alpha \in [0, N-2]$ *and* $\mathcal{H}^{\alpha}(E) = 0$. *If* f *is subharmonic in* $\Omega \setminus E$ *and satisfies*

$$\mathcal{A}(f^+, x, r) \leq C^* r^{\alpha+2-N} \quad (\overline{B^N(x,r)} \subset \Omega)$$

for some positive constant C^*, *then* f *has a subharmonic extension to* Ω.

Proof. As in the proof of Theorem 8.5, we may suppose that $\alpha \in (0, N-2)$ and $f \geq 0$. Since $f \in \mathcal{L}^1_{\mathrm{loc}}(\Omega)$, it is sufficient to show that

$$\int f \Delta\varphi \, dm_N \geq 0 \tag{8.8}$$

for any nonnegative testfunction $\varphi \in \mathcal{D}(\Omega)$. Take such a φ arbitrarily. And again as in the proof of Theorem 8.5, we may suppose that $0 \leq \varphi \leq 1$ and $|\mathcal{D}^{\lambda}\varphi| \leq 1$ for each multi-index λ, $|\lambda| \leq 2$. Let $K = \operatorname{spt}\varphi$. Choose r_0, $0 < r_0 < 1$, such that $\hat{K} = \overline{K}_{r_0} \subset K_{2r_0} \subset \overline{K}_{2r_0} \subset \Omega$. Let $\varepsilon > 0$ be arbitrarily given. We will cover K by a suitable collection of mutually disjoint dyadic cubes. This will be done in three steps.

First, using the assumption $\mathcal{H}^{\alpha}(E) = 0$ and (8.1), we find a sequence of mutually disjoint dyadic cubes Q_i, $s(Q_i) = s_i$, $i = 1, 2, \ldots$, such that

$$\sum_{i=1}^{+\infty} s_i^{\alpha} < \varepsilon. \tag{8.9}$$

We may suppose that $3 s_i \sqrt{N} < r_0$, $i = 1, 2, \ldots$. Since $E \cap \hat{K}$ is compact, there is $N_1 \in \mathbb{N}$ such that

$$E \cap \hat{K} \subset \cup_{i=1}^{N_1} Q_i. \tag{8.10}$$

Second, we attach to each cube Q_i, $s(Q_i) = s_i$, $i = 1, 2, \ldots, N_1$, all adjacent dyadic cubes with the same length s_i. Since two dyadic cubes are either mutually disjoint or one is contained in the other, we may drop extra cubes away. Proceeding in this way we get a collection of mutually disjoint cubes $Q_i^{j_i}$, $j_i = 0, 1, \ldots, n_i$, $i = 1, 2, \ldots, N_1$, such that

$$s(Q_i^{j_i}) = s(Q_i) = s_i, \quad j_i = 0, 1, \ldots, n_i \leq 3^N - 1, \; i = 1, 2, \ldots, N_1. \tag{8.11}$$

(That indeed $n_i \leq 3^N - 1$ for all $i = 1, 2, \ldots, N_1$, follows from the fact that we are considering adjacent cubes of the same length.)

Third, cover the remaining bounded set $K \setminus ((\cup_{i=1}^{N_1} Q_i) \cup (\cup_{i=1}^{N_1} (\cup_{j_i=0}^{n_i} Q_i^{j_i})))$ by mutually disjoint, dyadic cubes $\tilde{Q}_k$, all with the same length $s(\tilde{Q}_k) = s$, $k = 0, 1, \ldots, N_2$, where $s = \min\{ s_i : i = 1,2,\dots,N_1 \}$. Using then the facts that Q_i and $Q_i^{j_i}$ are adjacent, that $s(Q_i) = s(Q_i^{j_i}) = s_i$, $j_i = 0,1,\dots,n_i$, and $s(\tilde{Q}_k) = s \le s_i$, $i = 1,2,\dots,N_1$, $k = 0,1,\dots,N_2$, one sees easily that

$$\frac{3}{2}\tilde{Q}_k \cap E = \emptyset \text{ for } k = 0,1,\dots,N_2. \tag{8.12}$$

In order to show that (8.5) holds, we next choose nonnegative test functions φ_i, $\varphi_i^{j_i}$, $j_i = 0,1,\dots,n_i$, $i = 1,2,\dots,N_1$, and $\tilde{\varphi}_k$, $k = 0,1,\dots,N_2$, from $\mathcal{D}(\Omega)$ with the aid of the above Lemma, and thus with the following properties:

$$\operatorname{spt}\varphi_i \subset \frac{3}{2}Q_i,\ |\mathcal{D}^\lambda \varphi_i| \le \frac{C_\lambda}{s_i^{|\lambda|}} \text{ for } \lambda, |\lambda| \le 2,\ i = 1,2,\dots,N_1; \tag{8.13}$$

$$\operatorname{spt}\varphi_i^{j_i} \subset \frac{3}{2}Q_i^{j_i},\ |\mathcal{D}^\lambda \varphi_i^{j_i}| \le \frac{C_\lambda}{s_i^{|\lambda|}} \text{ for } \lambda, |\lambda| \le 2,\ j_i = 0,1,\dots,n_i;\ i = 1,2,\dots,N_1; \tag{8.14}$$

$$\operatorname{spt}\tilde{\varphi}_k \subset \frac{3}{2}\tilde{Q}_k,\ |\mathcal{D}^\lambda \tilde{\varphi}_k| \le \frac{C_\lambda}{s^{|\lambda|}} \text{ for } \lambda, |\lambda| \le 2,\ k = 0,1,\dots,N_2; \tag{8.15}$$

$$\sum_{i=1}^{N_1} \varphi_i(x) + \sum_{i=1}^{N_1}\sum_{j_i=0}^{n_i} \varphi_i^{j_i}(x) + \sum_{k=0}^{N_2} \tilde{\varphi}_k(x) = 1 \text{ for } x \in K. \tag{8.16}$$

Using then (8.9), (8.12) and the fact that f is subharmonic in $\Omega \setminus E$, one gets

$$\int_{\frac{3}{2}\tilde{Q}_k} f\,\Delta(\varphi\tilde{\varphi}_k)\,dm_N \ge 0 \text{ for } k = 0,1,\dots,N_2.$$

From this, (8.13), (8.10) and (8.11), it follows that

$$\begin{aligned}\int f\Delta\varphi\,dm_N &= \int f\Delta\left[\varphi\left(\sum_{i=1}^{N_1}\varphi_i + \sum_{i=1}^{N_1}\sum_{j_i=0}^{n_i}\varphi_i^{j_i} + \sum_{k=0}^{N_2}\tilde{\varphi}_k\right)\right] dm_N \ge \\ &\ge \sum_{i=1}^{N_1}\int_{\frac{3}{2}Q_i} f\,\Delta(\varphi\varphi_i)\,dm_N + \sum_{i=1}^{N_1}\sum_{j_i=0}^{n_i}\int_{\frac{3}{2}Q_i^{j_i}} f\,\Delta(\varphi\varphi_i^{j_i})\,dm_N.\end{aligned}$$

Using then (8.10) and (8.11) and proceeding then as in the proof of Theorem 8.5, we get similar estimates as in (8.3),

$$|\Delta(\varphi\varphi_i)(x)| \le \frac{C}{s_i^2} \text{ for } x \in \frac{3}{2}Q_i,\ i = 1,2,\dots,N_1,$$

and

$$|\Delta(\varphi\varphi_i^{j_i})(x)| \le \frac{C}{s_i^2} \text{ for } x \in \frac{3}{2}Q_i^{j_i},\ j_i = 0,1,\dots,n_i,\ i = 1,2,\dots,N_1.$$

In view of these inequalities, and of (8.10), (8.8) and (8.6), we get (in the sequel x_i and $x_i^{j_i}$ are the centers of the cubes Q_i, $Q_i^{j_i}$, $j_i = 0, 1, \ldots, n_i$, $i = 1, 2, \ldots, N_1$, respectively, and $\nu_N = m_N(B^N(0,1))$,

$$\int f \Delta\varphi \, dm_N \geq -C \left(\sum_{i=1}^{N_1} \frac{1}{s_i^2} \int_{\frac{3}{2}Q_i} f \, dm_N + \sum_{i=1}^{N_1} \sum_{j_i=0}^{n_i} \frac{1}{s_i^2} \int_{\frac{3}{2}Q_i^{j_i}} f \, dm_N \right) \geq$$

$$\geq -C \left(\sum_{i=1}^{N_1} \frac{1}{s_i^2} \int_{B^N(x_i, \frac{3}{4} s_i \sqrt{N})} f \, dm_N + \sum_{i=1}^{N_1} \sum_{j_i=0}^{n_i} \frac{1}{s_i^2} \int_{B^N(x_i^{j_i}, \frac{3}{4} s_i \sqrt{N})} f \, dm_N \right) \geq$$

$$\geq -\left(\frac{3}{4} \sqrt{N} \right)^N \nu_N C \left(\sum_{i=1}^{N_1} s_i^\alpha + 3^N \sum_{i=1}^{N_1} s_i^\alpha \right) \geq -C \sum_{i=1}^{N_1} s_i^\alpha \geq -C\varepsilon.$$

Since $C = C(N, \alpha, C^*)$ and ε was arbitrarily given, (8.8) follows and the proof is complete. □

9. Hausdorff Measure and Extension Results for Subharmonic Functions, for Separately Subharmonic Functions, for Harmonic Functions and for Separately Harmonic Functions

Abstract. We give extension results for subharmonic, separately subharmonic, harmonic and separately harmonic functions. Our results improve previous results of Blanchet.

Keywords. Subharmonic, separately subharmonic, harmonic, separately harmonic, Hausdorff measure

9.1. A result of Federer.

The following important result of Federer on geometric measure theory will be used repeatedly.

Lemma 9.1. ([35], Theorem 2.10.25, p. 188, and [135], Corollary 4, Lemma 2, p. 114) *Suppose that $E \subset \mathbb{R}^n$, $n \geq 2$. Let $\alpha \geq 0$ and let $\pi_k : \mathbb{R}^n \to \mathbb{R}^k$ denote the projection onto the first k coordinates.*

(i) *If $\mathcal{H}^{k+\alpha}(E) = 0$, then $\mathcal{H}^{\alpha}(E \cap \pi_k^{-1}\{x\}) = 0$ for $\mathcal{H}^k$-almost all $x \in \mathbb{R}^k$.*

(ii) *If $\mathcal{H}^{k+\alpha}(E) < +\infty$, then $\mathcal{H}^{\alpha}(E \cap \pi_k^{-1}\{x\}) < +\infty$ for $\mathcal{H}^k$-almost all $x \in \mathbb{R}^k$.*

9.2. A result of Blanchet.

Blanchet has given the following result:

Theorem 9.2. ([9],Theorems 3.1, 3.2 and 3.3, pp. 312-313) *Let Ω be a domain in $\mathbb{R}^n$, $n \geq 2$, and let S be a hypersurface of class $\mathcal{C}^1$ which divides Ω into two subdomains Ω_1 and Ω_2. Let $u \in \mathcal{C}^0(\Omega) \cap \mathcal{C}^2(\Omega_1 \cup \Omega_2)$ be subharmonic (respectively convex (or respectively plurisubharmonic provided Ω is then a domain in $\mathbb{C}^n$, $n \geq 1$)) in Ω_1 and Ω_2. If $u_i = u|\Omega_i \in \mathcal{C}^1(\Omega_i \cup S)$, $i = 1, 2$, and*

$$\frac{\partial u_i}{\partial \overline{n}^k} \geq \frac{\partial u_k}{\partial \overline{n}^k} \tag{9.1}$$

on S with $i, k = 1, 2$, then u is subharmonic (respectively convex (or respectively plurisubharmonic)) in Ω.

Above $\overline{n}^k = (\overline{n}_1^k, \overline{n}_2^k, \ldots, \overline{n}_n^k)$ is the unit normal exterior to Ω_k, and $u_k \in \mathcal{C}^1(\Omega_k \cup S)$, $k = 1, 2$, means that there exist n functions v_k^j, $j = 1, 2, \ldots, n$, continuous on $\Omega_k \cup S$, such that

$$v_k^j(x) = \frac{\partial u_k}{\partial x_j}(x)$$

for all $x \in \Omega_k$, $k = 1, 2$ and $j = 1, 2, \ldots, n$. The following example shows that one cannot drop the above condition (9.1) in Blanchet's theorem, see also [119, 121].

Example 18. The function $u : \mathbb{R}^2 \to \mathbb{R}$,

$$u(z) = u(x+iy) = u(x,y) := \begin{cases} 1+x, & \text{when } x < 0, \\ 1-x, & \text{when } x \geq 0, \end{cases}$$

is continuous in $\mathbb{R}^2$ and subharmonic, even harmonic in $\mathbb{R}^2 \setminus (\{0\} \times \mathbb{R})$. It is easy to see that u does not satisfy the condition (9.1) on $S = \{0\} \times \mathbb{R}$ and that u is not subharmonic in $\mathbb{R}^2$.

Juhani Riihentaus

Remark 9.3. For related results, previous and later, see Khabibullin's results [58], Lemma 2.2, p. 201, Fundamental Theorem 2.1, pp. 200-201, and [59], Lemma 4.1, p. 503, Theorem 2.1, p. 498, Theorems 3.1 and 3.2, pp. 500-501. In this connection, see also [46], 1.4.3, pp. 21-22.

9.3. **An improvement to the subharmonic extension result of Blanchet.** Already in [102], Theorem 4, pp. 181-182, we have given partial improvements to the cited subharmonic removability results of Blanchet. For other previous improvements, see [119], Theorem, p. 568, and [121], Theorem 1, p. 154. Instead of hypersurfaces of class $\mathcal{C}^1$, we have there considered arbitrary sets of finite $(n-1)$-dimensional Hausdorff measure as exceptional sets. However, now we recall, and use, our most recent result, [122], Theorem 1, p. 61, see Theorem 9.4 below, where Blanchet's condition (9.1) is replaced by the condition (iv), which is now, at least seemingly, less stringent than in our previous results [119], Theorem, (iv), p. 568, and [121], Theorem 1, (iv), p. 154, say.

Theorem 9.4. ([122], Theorem 1, p. 61) *Suppose that Ω is a domain in $\mathbb{R}^n$, $n \geq 2$. Let $E \subset \Omega$ be closed in Ω and let $\mathcal{H}^{n-1}(E) < +\infty$. Let $u : \Omega \setminus E \to \mathbb{R}$ be subharmonic and such that the following conditions are satisfied:*

(i) $u \in \mathcal{L}^1_{\mathrm{loc}}(\Omega)$.

(ii) $u \in \mathcal{C}^2(\Omega \setminus E)$.

(iii) *For each j, $1 \leq j \leq n$, $\frac{\partial^2 u}{\partial x_j^2} \in \mathcal{L}^1_{\mathrm{loc}}(\Omega)$.*

(iv) *For each j, $1 \leq j \leq n$, and for $\mathcal{H}^{n-1}$-almost all $X_j \in \mathbb{R}^{n-1}$ such that $E(X_j)$ is finite, the following condition holds:*
For each $x_j^0 \in E(X_j)$ there exist sequences $x_{j,l}^{0,1}, x_{j,l}^{0,2} \in (\Omega \setminus E)(X_j)$, $l = 1, 2, \ldots$, such that $x_{j,l}^{0,1} \nearrow x_j^0$, $x_{j,l}^{0,2} \searrow x_j^0$ as $l \to +\infty$, and

(iv(a)) $\lim_{l \to +\infty} u(x_{j,l}^{0,1}, X_j) = \lim_{l \to +\infty} u(x_{j,l}^{0,2}, X_j) \in \mathbb{R}$,

(iv(b)) $-\infty < \liminf_{l \to +\infty} \frac{\partial u}{\partial x_j}(x_{j,l}^{0,1}, X_j) \leq \limsup_{l \to +\infty} \frac{\partial u}{\partial x_j}(x_{j,l}^{0,2}, X_j) < +\infty$.

Then u has a subharmonic extension to Ω.

Proof. Observe first that using suitable subsequences one can replace the assumption (iv) by the following, (only) seemingly stronger condition:

(iv*) *For each $x_j^0 \in E(X_j)$ there exist sequences $x_{j,l}^{0,1}, x_{j,l}^{0,2} \in (\Omega \setminus E)(X_j)$, $l = 1, 2, \ldots$, such that $x_{j,l}^{0,1} \nearrow x_j^0$, $x_{j,l}^{0,2} \searrow x_j^0$ as $l \to +\infty$, and*

(iv*(a)) $\lim_{l \to +\infty} u(x_{j,l}^{0,1}, X_j) = \lim_{l \to +\infty} u(x_{j,l}^{0,2}, X_j) \in \mathbb{R}$,

(iv*(b)) $-\infty < \lim_{l \to +\infty} \frac{\partial u}{\partial x_j}(x_{j,l}^{0,1}, X_j) \leq \lim_{l \to +\infty} \frac{\partial u}{\partial x_j}(x_{j,l}^{0,2}, X_j) < +\infty$.

It is sufficient to show that

$$\int u(x)\, \Delta\varphi(x)\, dx \geq 0$$

for all nonnegative testfunctions $\varphi \in \mathcal{D}(\Omega)$, see e.g. [49], Corollary 1, p. 13.

Take $\varphi \in \mathcal{D}(\Omega)$, $\varphi \geq 0$, arbitrarily. Let $K = \operatorname{spt}\varphi$. Choose a domain Ω_1 such that $K \subset \Omega_1 \subset \overline{\Omega}_1 \subset \Omega$ and $\overline{\Omega}_1$ is compact. Since $u \in \mathcal{C}^2(\Omega \setminus E)$ and u is subharmonic in $\Omega \setminus E$, $\Delta u(x) \geq 0$ for all $x \in \Omega \setminus E$. Thus the claim follows if we show that

$$\int u(x)\, \Delta\varphi(x)\, dx \geq \int \Delta u(x)\, \varphi(x)\, dx.$$

For this purpose fix j, $1 \leq j \leq n$, arbitrarily for a while. By Fubini's theorem, see e.g. [33], Theorem 1, pp. 22-23,

$$\int u(x) \frac{\partial^2 \varphi}{\partial x_j^2}(x)\, dx = \int \Big[\int u(x_j, X_j) \frac{\partial^2 \varphi}{\partial x_j^2}(x_j, X_j)\, dx_j\Big]\, dX_j.$$

Using the above Lemma, assumptions (i), (ii) and (iii), and Fubini's theorem, we see that for $\mathcal{H}^{n-1}$-almost all $X_j \in \mathbb{R}^{n-1}$,

$$\text{(9.2)} \qquad \begin{cases} u(\,\cdot\,, X_j) \in \mathcal{L}^1_{\mathrm{loc}}(\Omega(X_j)), \\ \dfrac{\partial^2 u}{\partial x_j^2}(\,\cdot\,, X_j) \in \mathcal{L}^1_{\mathrm{loc}}(\Omega(X_j)), \\ E(X_j) \ \text{ is finite, thus there exists } M = M(X_j) \in \mathbb{N}_0 \text{ such that} \\ E(X_j) = \{x_j^1, x_j^2, \ldots, x_j^M\} \text{ where } x_j^k < x_j^{k+1},\ k = 1, 2, \ldots, M-1. \end{cases}$$

Let $X_j \in \mathbb{R}^{n-1}$ be arbitrary as above in (9.2). We may suppose that $\Omega(X_j)$ is a finite interval. Choose for each $k = 1, 2, \ldots, M$ numbers $a_k, b_k \in (\Omega \setminus E)(X_j)$ such that $a_k < x_j^k < b_k$, $k = 1, 2, \ldots, M$, $a_{k+1} = b_k$, $k = 1, 2, \ldots, M-1$, and that $a_1, b_M \in (\Omega \setminus \overline{\Omega}_1)(X_j)$.

With the aid of (iv*) we find for each $x_j^k \in E(X_j)$ sequences $x_{j,l}^{k,1}, x_{j,l}^{k,2} \in (\Omega \setminus E)(X_j)$, $l = 1, 2, \ldots$, $x_{j,l}^{k,1} \nearrow x_j^k$, $x_{j,l}^{k,2} \searrow x_j^k$ as $l \to +\infty$, such that

$$\lim_{l \to +\infty} u(x_{j,l}^{k,1}, X_j) = \lim_{l \to +\infty} u(x_{j,l}^{k,2}, X_j) \in \mathbb{R},$$

and

$$-\infty < \lim_{l \to +\infty} \frac{\partial u}{\partial x_j}(x_{j,l}^{k,1}, X_j) \leq \lim_{l \to +\infty} \frac{\partial u}{\partial x_j}(x_{j,l}^{k,2}, X_j) < +\infty.$$

Take k, $1 \leq k \leq M$, arbitrarily and consider the interval (a_k, b_k), where $a_k < x_j^k < b_k$. To simplify the notation, write $a := a_k$, $b := b_k$ and $x_j^0 := x_j^k$. Then

$$a < x_{j,l}^{0,1} \nearrow x_j^0,\ b > x_{j,l}^{0,2} \searrow x_j^0 \text{ as } l \to +\infty.$$

Using partial integration we get:

$$\int_a^b u(x_j,X_j)\,\frac{\partial^2\varphi}{\partial x_j^2}(x_j,X_j)\,dx_j = \int_a^{x_j^0} u(x_j,X_j)\,\frac{\partial^2\varphi}{\partial x_j^2}(x_j,X_j)\,dx_j +$$

$$+\int_{x_j^0}^b u(x_j,X_j)\,\frac{\partial^2\varphi}{\partial x_j^2}(x_j,X_j)\,dx_j =$$

$$= \lim_{l\to+\infty}\int_a^{x_{j,l}^{0,1}} u(x_j,X_j)\,\frac{\partial^2\varphi}{\partial x_j^2}(x_j,X_j)\,dx_j + \lim_{l\to+\infty}\int_{x_{j,l}^{0,2}}^b u(x_j,X_j)\,\frac{\partial^2\varphi}{\partial x_j^2}(x_j,X_j)\,dx_j =$$

$$= \lim_{l\to+\infty}\left[\Big|_a^{x_{j,l}^{0,1}} u(x_j,X_j)\,\frac{\partial\varphi}{\partial x_j}(x_j,X_j) - \int_a^{x_{j,l}^{0,1}} \frac{\partial u}{\partial x_j}(x_j,X_j)\,\frac{\partial\varphi}{\partial x_j}(x_j,X_j)\,dx_j\right] +$$

$$+\lim_{l\to+\infty}\left[\Big|_{x_{j,l}^{0,2}}^b u(x_j,X_j)\frac{\partial\varphi}{\partial x_j}(x_j,X_j) - \int_{x_{j,l}^{0,2}}^b \frac{\partial u}{\partial x_j}(x_j,X_j)\,\frac{\partial\varphi}{\partial x_j}(x_j,X_j)\,dx_j\right] =$$

$$= \left[u(b,X_j)\frac{\partial\varphi}{\partial x_j}(b,X_j) - u(a,X_j)\frac{\partial\varphi}{\partial x_j}(a,X_j)\right] +$$

$$+\lim_{l\to+\infty}\left[u(x_{j,l}^{0,1},X_j)\,\frac{\partial\varphi}{\partial x_j}(x_{j,l}^{0,1},X_j) - \int_a^{x_{j,l}^{0,1}} \frac{\partial u}{\partial x_j}(x_j,X_j)\,\frac{\partial\varphi}{\partial x_j}(x_j,X_j)\,dx_j\right] +$$

$$-\lim_{l\to+\infty}\left[u(x_{j,l}^{0,2},X_j)\,\frac{\partial\varphi}{\partial x_j}(x_{j,l}^{0,2},X_j) + \int_{x_{j,l}^{0,2}}^b \frac{\partial u}{\partial x_j}(x_j,X_j)\,\frac{\partial\varphi}{\partial x_j}(x_j,X_j)\,dx_j\right] =$$

$$= \left[u(b,X_j)\frac{\partial\varphi}{\partial x_j}(b,X_j) - u(a,X_j)\frac{\partial\varphi}{\partial x_j}(a,X_j)\right] +$$

$$-\lim_{l\to+\infty}\int_a^{x_{j,l}^{0,1}} \frac{\partial u}{\partial x_j}(x_j,X_j)\,\frac{\partial\varphi}{\partial x_j}(x_j,X_j)\,dx_j - \lim_{l\to+\infty}\int_{x_{j,l}^{0,2}}^b \frac{\partial u}{\partial x_j}(x_j,X_j)\,\frac{\partial\varphi}{\partial x_j}(x_j,X_j)\,dx_j =$$

$$= \left[u(b,X_j)\frac{\partial\varphi}{\partial x_j}(b,X_j) - u(a,X_j)\frac{\partial\varphi}{\partial x_j}(a,X_j)\right] +$$

$$-\lim_{l\to+\infty}\left[\Big|_a^{x_{j,l}^{0,1}} \frac{\partial u}{\partial x_j}(x_j,X_j)\varphi(x_j,X_j) - \int_a^{x_{j,l}^{0,1}} \frac{\partial^2 u}{\partial x_j^2}(x_j,X_j)\,\varphi(x_j,X_j)\,dx_j\right] +$$

$$-\lim_{l\to+\infty}\left[\Big|_{x_{j,l}^{0,2}}^b \frac{\partial u}{\partial x_j}(x_j,X_j)\varphi(x_j,X_j) - \int_{x_{j,l}^{0,2}}^b \frac{\partial^2 u}{\partial x_j^2}(x_j,X_j)\,\varphi(x_j,X_j)\,dx_j\right] =$$

$$= \left[u(b,X_j)\frac{\partial\varphi}{\partial x_j}(b,X_j) - u(a,X_j)\frac{\partial\varphi}{\partial x_j}(a,X_j)\right] +$$

$$+\left[\frac{\partial u}{\partial x_j}(a,X_j)\varphi(a,X_j) - \frac{\partial u}{\partial x_j}(b,X_j)\varphi(b,X_j)\right] +$$

$$-\lim_{l\to+\infty}\left[\frac{\partial u}{\partial x_j}(x_{j,l}^{0,1},X_j)\varphi(x_{j,l}^{0,1},X_j)\right] +$$

$$+\lim_{l\to+\infty}\left[\frac{\partial u}{\partial x_j}(x_{j,l}^{0,2},X_j)\varphi(x_{j,l}^{0,2},X_j)\right] + \int_a^b \frac{\partial^2 u}{\partial x_j^2}(x_j,X_j)\,\varphi(x_j,X_j)\,dx_j =$$

$$= \left[u(b,X_j)\frac{\partial\varphi}{\partial x_j}(b,X_j) - u(a,X_j)\frac{\partial\varphi}{\partial x_j}(a,X_j)\right] +$$
$$+ \left[\frac{\partial u}{\partial x_j}(a,X_j)\varphi(a,X_j) - \frac{\partial u}{\partial x_j}(b,X_j)\varphi(b,X_j)\right] +$$
$$+ \left[\lim_{l\to+\infty}\frac{\partial u}{\partial x_j}(x_{j,l}^{0,2},X_j) - \lim_{l\to+\infty}\frac{\partial u}{\partial x_j}(x_{j,l}^{0,1},X_j)\right]\varphi(x_j^0,X_j) +$$
$$+ \int_a^b \frac{\partial^2 u}{\partial x_j^2}(x_j,X_j)\,\varphi(x_j,X_j)\,dx_j \geq$$
$$\geq \left[u(b,X_j)\frac{\partial\varphi}{\partial x_j}(b,X_j) - u(a,X_j)\frac{\partial\varphi}{\partial x_j}(a,X_j)\right] +$$
$$+ \left[\frac{\partial u}{\partial x_j}(a,X_j)\varphi(a,X_j) - \frac{\partial u}{\partial x_j}(b,X_j)\varphi(b,X_j)\right] + \int_a^b \frac{\partial^2 u}{\partial x_j^2}(x_j,X_j)\,\varphi(x_j,X_j)\,dx_j.$$

Above we have used just standard properties of limits and our assumption (iv*(b)).

To return to our original notation, we have thus obtained for each $k = 1,2,\ldots,M$,

$$\int_{a_k}^{b_k} u(x_j,X_j)\,\frac{\partial^2\varphi}{\partial x_j^2}(x_j,X_j)\,dx_j \geq \left[u(b_k,X_j)\frac{\partial\varphi}{\partial x_j}(b_k,X_j) - u(a_k,X_j)\frac{\partial\varphi}{\partial x_j}(a_k,X_j)\right] +$$
$$+ \left[\frac{\partial u}{\partial x_j}(a_k,X_j)\varphi(a_k,X_j) - \frac{\partial u}{\partial x_j}(b_k,X_j)\varphi(b_k,X_j)\right] + \int_{a_k}^{b_k} \frac{\partial^2 u}{\partial x_j^2}(x_j,X_j)\,\varphi(x_j,X_j)\,dx_j.$$

Then just sum over $k = 1,2,\ldots,M$:

$$\int u(x_j,X_j)\,\frac{\partial^2\varphi}{\partial x_j^2}(x_j,X_j)\,dx_j = \int_{a_1}^{b_M} u(x_j,X_j)\,\frac{\partial^2\varphi}{\partial x_j^2}(x_j,X_j)\,dx_j =$$
$$= \sum_{k=1}^{M}\int_{a_k}^{b_k} u(x_j,X_j)\,\frac{\partial^2\varphi}{\partial x_j^2}(x_j,X_j)\,dx_j \geq$$
$$\geq \sum_{k=1}^{M}\left[u(b_k,X_j)\frac{\partial\varphi}{\partial x_j}(b_k,X_j) - u(a_k,X_j)\frac{\partial\varphi}{\partial x_j}(a_k,X_j)\right] +$$
$$+ \sum_{k=1}^{M}\left[\frac{\partial u}{\partial x_j}(a_k,X_j)\varphi(a_k,X_j) - \frac{\partial u}{\partial x_j}(b_k,X_j)\varphi(b_k,X_j)\right] +$$
$$+ \sum_{k=1}^{M}\int_{a_k}^{b_k} \frac{\partial^2 u}{\partial x_j^2}(x_j,X_j)\,\varphi(x_j,X_j)\,dx_j =$$
$$= \int_{a_1}^{b_M} \frac{\partial^2 u}{\partial x_j^2}(x_j,X_j)\,\varphi(x_j,X_j)\,dx_j = \int \frac{\partial^2 u}{\partial x_j^2}(x_j,X_j)\,\varphi(x_j,X_j)\,dx_j.$$

Above we have used the choice of the numbers a_k, b_k, $k = 1,2,\ldots,M$, and the fact that $a_1, b_M \in (\Omega\setminus\overline{\Omega}_1)(X_j)$.

Integrate then with respect to X_j and use again Fubini's theorem:

$$\int u(x)\frac{\partial^2\varphi}{\partial x_j^2}(x)\,dx = \int\left[\int u(x_j,X_j)\frac{\partial^2\varphi}{\partial x_j^2}(x_j,X_j)\,dx_j\right]dX_j \geq$$

$$\geq \int\left[\int \frac{\partial^2 u}{\partial x_j^2}(x_j,X_j)\varphi(x_j,X_j)\,dx_j\right]dX_j = \int \frac{\partial^2 u}{\partial x_j^2}(x)\varphi(x)\,dx.$$

Summing over $j = 1,2,\ldots,n$ gives the desired inequality

$$\int u(x)\Delta\varphi(x)\,dx = \int u(x)\sum_{j=1}^{n}\frac{\partial^2\varphi}{\partial x_j^2}(x)\,dx \geq$$

$$\geq \int \sum_{j=1}^{n}\frac{\partial^2 u}{\partial x_j^2}(x)\varphi(x)\,dx = \int \Delta u(x)\varphi(x)\,dx \geq 0,$$

concluding the proof. □

Example 19. The function $u : \mathbb{R}^2 \to \mathbb{R}$, given already in Example 18, is continuous in $\mathbb{R}^2$ and subharmonic, even harmonic in $\mathbb{R}^2 \setminus (\{0\} \times \mathbb{R})$, but not subharmonic in $\mathbb{R}^2$. Observe that u satisfies the above conditions (i), (ii), (iii) and (iv(a)) in $\mathbb{R}^2 \setminus (\{0\} \times \mathbb{R})$. However, $u|\mathbb{R}^2 \setminus (\{0\} \times \mathbb{R})$ does not satisfy the condition (iv(b)). Thus one cannot drop the condition (iv(b)) in Theorem 9.4.

The following corollary reflects the strength of Theorem 9.4.

Corollary 9.5. ([122], Corollary 1, p. 61) *Suppose that Ω is a domain in $\mathbb{R}^n$, $n \geq 2$. Let $E \subset \Omega$ be closed in Ω and let $\mathcal{H}^{n-1}(E) = 0$. Let $u : \Omega \setminus E \to \mathbb{R}$ be subharmonic and such that the following conditions are satisfied:*

(i) $u \in \mathcal{L}^1_{\text{loc}}(\Omega)$,
(ii) $u \in \mathcal{C}^2(\Omega \setminus E)$,
(iii) *for each* j, $1 \leq j \leq n$, $\dfrac{\partial^2 u}{\partial x_j^2} \in \mathcal{L}^1_{\text{loc}}(\Omega)$.

Then u has a subharmonic extension to Ω.

Proof. Follows directly from Theorem 9.4 and from the above Lemma of Federer. □

9.4. An extension result for separately subharmonic functions.

9.4.1. Next we will give an extension result for separately subharmonic functions. Our proof will be based on Theorem 9.4 and on the following nice result, Proposition 9.6 below. Observe that the there used *hypoharmonic functions* are in our terminology just *subharmonic functions*.

Proposition 9.6. ([49], Proposition 1, p. 33) *Suppose that Ω is a domain in $\mathbb{R}^{p+q}$, $p,q \geq 2$. Let $w : \Omega \to [-\infty,+\infty)$ be nearly subharmonic. Let $w^* : \Omega \to [-\infty,+\infty)$ be the regularized function of w, which is then subharmonic. Then the following properties are equivalent.*

(1) *The distribution* $\Delta_x w = \Delta_x w^* =$ *(sum of the square second order derivatives of* w *or* w^* *with respect to the* p *coordinates of* x*) is positive.*

(2) *For all* $y \in \mathbb{R}^q$ *the function*

$$\Omega(y) \ni x \mapsto w^*(x,y) \in [-\infty, +\infty)$$

is hypoharmonic.

(3) *For almost every* $y \in \mathbb{R}^q$ *the function*

$$\Omega(y) \ni x \mapsto w^*(x,y) \in [-\infty, +\infty)$$

is subharmonic.

(4) *For almost every* $y \in \mathbb{R}^q$ *the function*

$$\Omega(y) \ni x \mapsto w(x,y) \in [-\infty, +\infty)$$

is nearly subharmonic.

Then our result. Observe that our proof below is a slightly improved version of our original proof in [122], pp. 62-64.

Theorem 9.7. ([122], Theorem 2, p. 62) *Suppose that* Ω *is a domain in* $\mathbb{R}^{p+q}$, $p, q \geq 2$. *Let* $E \subset \Omega$ *be closed in* Ω *and let* $\mathcal{H}^{p+q-1}(E) < +\infty$. *Let* $w : \Omega \setminus E \to \mathbb{R}$ *be separately subharmonic, that is,*

for all $y \in \mathbb{R}^q$ *the function* $(\Omega \setminus E)(y) \ni x \mapsto w(x,y) \in \mathbb{R}$ *is subharmonic,*

and

for all $x \in \mathbb{R}^p$ *the function* $(\Omega \setminus E)(x) \ni y \mapsto w(x,y) \in \mathbb{R}$ *is subharmonic,*

and such that the following conditions are satisfied:

(i) $w \in \mathcal{L}^1_{\mathrm{loc}}(\Omega)$.

(ii) $w \in \mathcal{C}^2(\Omega \setminus E)$.

(iii) *For each* j, $1 \leq j \leq p$, $\frac{\partial^2 w}{\partial x_j^2} \in \mathcal{L}^1_{\mathrm{loc}}(\Omega)$, *and for each* k, $1 \leq k \leq q$, $\frac{\partial^2 w}{\partial y_k^2} \in \mathcal{L}^1_{\mathrm{loc}}(\Omega)$.

(iv) *For each* j, $1 \leq j \leq p$, *and for* $\mathcal{H}^{p-1+q}$*-almost all* $(X_j, y) \in \mathbb{R}^{p-1+q}$ *such that* $E(X_j, y)$ *is finite, the following condition holds:*
For each $x_j^0 \in E(X_j, y)$ *there exist sequences* $x_{j,l}^{0,1}, x_{j,l}^{0,2} \in (\Omega \setminus E)(X_j, y)$, $l = 1, 2, \ldots$, *such that* $x_{j,l}^{0,1} \nearrow x_j^0$, $x_{j,l}^{0,2} \searrow x_j^0$ *as* $l \to +\infty$, *and*

(iv(a)) $\lim_{l \to +\infty} w(x_{j,l}^{0,1}, X_j, y) = \lim_{l \to +\infty} w(x_{j,l}^{0,2}, X_j, y) \in \mathbb{R}$,

(iv(b)) $-\infty < \liminf_{l \to +\infty} \frac{\partial w}{\partial x_j}(x_{j,l}^{0,1}, X_j, y) \leq \limsup_{l \to +\infty} \frac{\partial w}{\partial x_j}(x_{j,l}^{0,2}, X_j, y) < +\infty$.

(v) *For each* k, $1 \leq k \leq q$, *and for* $\mathcal{H}^{p+q-1}$*-almost all* $(x, Y_k) \in \mathbb{R}^{p+q-1}$ *such that* $E(x, Y_k)$ *is finite, the following condition holds:*
For each $y_k^0 \in E(x, Y_k)$ *there exist sequences* $y_{k,l}^{0,1}, y_{k,l}^{0,2} \in (\Omega \setminus E)(x, Y_k)$, $l = 1, 2, \ldots$, *such that* $y_{k,l}^{0,1} \nearrow y_k^0$, $y_{k,l}^{0,2} \searrow y_k^0$ *as* $l \to +\infty$, *and*

(v(a)) $\lim_{l \to +\infty} w(x, y_{k,l}^{0,1}, Y_k) = \lim_{l \to +\infty} w(x, y_{k,l}^{0,2}, Y_k) \in \mathbb{R}$,

(v(b)) $-\infty < \liminf_{l \to +\infty} \frac{\partial w}{\partial y_k}(x, y_{k,l}^{0,1}, Y_k) \leq \limsup_{l \to +\infty} \frac{\partial w}{\partial y_k}(x, y_{k,l}^{0,2}, Y_k) < +\infty$.

Then w has a separately subharmonic extension to Ω.

Proof. Since w is separately subharmonic in $\Omega \setminus E$ and $w \in \mathcal{L}^1_{\mathrm{loc}}(\Omega)$, it follows from [109], Theorem A, p. 50, or from [116], Corollary 4.6, p. 412, say, that w is subharmonic in $\Omega \setminus E$. Using then Theorem 9.4 one sees that $w : \Omega \setminus E \to \mathbb{R}$ has a subharmonic extension $w^* : \Omega \to [-\infty, +\infty)$.

To show that the subharmonic function $w^* : \Omega \to [-\infty, +\infty)$ is in fact separately subharmonic, we must show that for all $y \in \mathbb{R}^q$ the function

$$\Omega(y) \ni x \mapsto w^*(x,y) \in [-\infty, +\infty)$$

is subharmonic, and that for all $x \in \mathbb{R}^p$ the function

$$\Omega(x) \ni y \mapsto w^*(x,y) \in [-\infty, +\infty)$$

is subharmonic. It is clearly sufficient to show that the first claim holds.

For this purpose we define $\tilde{w} : \Omega \to [-\infty, +\infty)$,

$$\tilde{w}(x,y) := \begin{cases} w(x,y), & \text{when } (x,y) \in \Omega \setminus E, \\ -\infty, & \text{when } (x,y) \in E. \end{cases}$$

To see that $\tilde{w}$ is nearly subharmonic, observe first that $\tilde{w}(x,y) = w(x,y)$ for all $(x,y) \in \Omega \setminus E$, thus for almost all $(x,y) \in \Omega$. Hence $\tilde{w} \in \mathcal{L}^1_{\mathrm{loc}}(\Omega)$. To see that the mean value inequality holds, take $(x_0, y_0) \in \Omega$ and $\overline{B^{p+q}((x_0,y_0),r)} \subset \Omega$ arbitrarily. If $(x_0, y_0) \in \Omega \setminus E$, then

$$\begin{aligned} \tilde{w}(x_0,y_0) = w(x_0,y_0) = w^*(x_0,y_0) \leq \\ \leq \frac{1}{\nu_{p+q} r^{p+q}} \int_{B^{p+q}((x_0,y_0),r)} w^*(x,y)\, dm_{p+q}(x,y) = \\ \leq \frac{1}{\nu_{p+q} r^{p+q}} \int_{B^{p+q}((x_0,y_0),r)} w(x,y)\, dm_{p+q}(x,y) = \\ \leq \frac{1}{\nu_{p+q} r^{p+q}} \int_{B^{p+q}((x_0,y_0),r)} \tilde{w}(x,y)\, dm_{p+q}(x,y). \end{aligned}$$

If $(x_0, y_0) \in E$, then, since $\tilde{w} \in \mathcal{L}^1_{\mathrm{loc}}(\Omega)$,

$$-\infty = \tilde{w}(x_0,y_0) < \tfrac{1}{\nu_{p+q} r^{p+q}} \textstyle\int_{B^{p+q}((x_0,y_0),r)} \tilde{w}(x,y)\, dm_{p+q}(x,y) \in \mathbb{R}.$$

Thus $\tilde{w}$ is nearly subharmonic.

In order to be able to use Proposition 9.6, we show that for $\mathcal{H}^q$-almost all $y \in \mathbb{R}^q$, the function

$$\Omega(y) \ni x \mapsto \tilde{w}(x,y) \in [-\infty, +\infty)$$

is nearly subharmonic.

For this purpose fix j, $1 \leq j \leq p$, arbitrarily for a while.

By our assumption $\mathcal{H}^{p-1+q}(E) < +\infty$. From the above Lemma of Federer it follows that for $\mathcal{H}^{p-1+q}$-almost all $(X_j, y) \in \mathbb{R}^{p-1+q}$ the set $E(y)(X_j)$ is finite. Write

$$A_j := \{(X_j, y) \in \mathbb{R}^{p-1+q} : E(y)(X_j) \text{ is finite}\}.$$

Thus

$$\mathcal{H}^{p-1+q}(A_j^c) = 0 \Leftrightarrow m_{p-1+q}(A_j^c) = 0 \Leftrightarrow \int_{\mathbb{R}^{p-1+q}} \chi_{A_j^c}(X_j, y) dm_{p-1+q}(X_j, y) = 0,$$

where $\chi_{A_j^c}(\cdot, \cdot)$ is the characteristic function of the set A_j^c, the complement taken in $\mathbb{R}^{p-1+q}$.

Next use Fubini's theorem:

$$0 = \int_{\mathbb{R}^{p-1+q}} \chi_{A_j^c}(X_j, y) dm_{p-1+q}(X_j, y) = \int_{\mathbb{R}^q} [\int_{\mathbb{R}^{p-1}} \chi_{A_j^c}(X_j, y) dm_{p-1}(X_j)] dm_q(y).$$

Since

$$\int_{\mathbb{R}^{p-1}} \chi_{A_j^c}(X_j, y) dm_{p-1}(X_j) \geq 0,$$

we see that in fact

$$\int_{\mathbb{R}^{p-1}} \chi_{A_j^c}(X_j, y) dm_{p-1}(X_j) = 0$$

for $\mathcal{H}^q$-almost all $y \in \mathbb{R}^q$.

Write

$$\begin{aligned}
B_1^j : &= \{y \in \mathbb{R}^q : \int_{\mathbb{R}^{p-1}} \chi_{A_j^c}(X_j, y) dm_{p-1}(X_j) = 0\} = \\
&= \{y \in \mathbb{R}^q : \chi_{A_j^c}(X_j, y) = 0 \text{ for } \mathcal{H}^{p-1}-\text{almost all } X_j \in \mathbb{R}^{p-1}\} = \\
&= \{y \in \mathbb{R}^q : \chi_{A_j}(X_j, y) = 1 \text{ for } \mathcal{H}^{p-1}-\text{almost all } X_j \in \mathbb{R}^{p-1}\} = \\
&= \{y \in \mathbb{R}^q : E(y)(X_j) \text{ is finite for } \mathcal{H}^{p-1}-\text{almost all } X_j \in \mathbb{R}^{p-1}\}.
\end{aligned}$$

Write $B_1 := B_1^1 \cap B_1^2 \cap \cdots \cap B_1^p$. Then for all $y \in B_1$ we have $(X_j, y) \in A_j$, that is, $E(y)(X_j)$ is finite for $\mathcal{H}^{p-1}$-almost all $X_j \in \mathbb{R}^{p-1}$, and this for all $j = 1, 2, \ldots, p$.

Next write

$$\begin{aligned}
B_2 &:= \{y \in \mathbb{R}^q : \mathcal{H}^{p-1}(E(y)) < +\infty\}, \\
B_3 &:= \{y \in \mathbb{R}^q : w(\cdot, y) \in \mathcal{L}^1_{\text{loc}}(\Omega(y))\}, \\
B_4 &:= \{y \in \mathbb{R}^q : w(\cdot, y) \in \mathcal{C}^2((\Omega \setminus E)(y))\}, \\
B_5^j &:= \{y \in \mathbb{R}^q : \frac{\partial^2}{\partial x_j^2} w(\cdot, y) \in \mathcal{L}^1_{\text{loc}}(\Omega(y))\}, \\
B_5 &:= B_5^1 \cap B_5^2 \cap \cdots \cap B_5^p, \\
B : &= B_1 \cap B_2 \cap B_3 \cap B_4 \cap B_5.
\end{aligned}$$

Then for all $y \in B$ the function

$$(\Omega \setminus E)(y) \ni x \mapsto w(x, y) \in \mathbb{R}$$

satisfies the assumptions of Theorem 9.4. Therefore these functions have nearly subharmonic extensions

$$\Omega(y) \ni x \mapsto w^{**}(x,y) \in [-\infty,+\infty).$$

To complete the proof, we show that for all $y \in B$ the function

$$\Omega(y) \ni x \mapsto \tilde{w}(x,y) \in [-\infty,+\infty)$$

is nearly subharmonic. Observe first that $\tilde{w}(\cdot,y) \in \mathcal{L}^1_{\mathrm{loc}}(\Omega(y))$, since $y \in B$. To show that $\tilde{w}(\cdot,y)$, $y \in B$, satisfies the mean value inequality, take $x_0 \in (\Omega \setminus E)(y)$ arbitrarily. Since $\mathcal{H}^{p-1}(E(y)) < +\infty$, we have

$$\begin{aligned}\tilde{w}(x_0,y) =& w(x_0,y) = w^{**}(x_0,y) \leq \frac{1}{\nu_p r^p}\int_{B^p(x_0,r)} w^{**}(x,y)\,dm_p(x) = \\ \leq& \frac{1}{\nu_p r^p}\int_{B^p(x_0,r)} w(x,y)\,dm_p(x) = \frac{1}{\nu_p r^p}\int_{B^p(x_0,r)} \tilde{w}(x,y) dm_p(x).\end{aligned}$$

In the case $x_0 \in E(y)$ we have $\tilde{w}(x_0,y) = -\infty$, and thus the mean value inequality is automatically satisfied.

Our claim follows now from Proposition 9.6. □

Example 20. ([122], Example 1, p. 64) The function $u : \mathbb{R}^4 \to \mathbb{R}$,

$$u(z_1,z_2) = u(x_1+iy_1,x_2+iy_2) = u(x_1,y_1,x_2,y_2) := \begin{cases} 1+x_1, & \text{when } x_1 < 0, \\ 1-x_1, & \text{when } x_1 \geq 0, \end{cases}$$

is continuous in $\mathbb{R}^4$ and separately subharmonic, even separately harmonic in $\mathbb{R}^4 \setminus (\{0\} \times \mathbb{R}^3)$, but not separately subharmonic in $\mathbb{R}^4$. As a matter of fact, it is easy to see that u is not even subharmonic in $\mathbb{R}^4$. Observe that u satisfies the above conditions (i), (ii), (iii), (iv(a)) and (v(a)) in $\mathbb{R}^4 \setminus (\{0\} \times \mathbb{R}^3)$. However, $u|\mathbb{R}^4 \setminus (\{0\} \times \mathbb{R}^3)$ does not satisfy the conditions (iv(b)) and (v(b)). Thus these conditions cannot be dropped in Theorem 9.7.

Corollary 9.8. ([122], Corollary 2, p. 64) *Suppose that Ω is a domain in $\mathbb{R}^{p+q}$, $p,q \geq 2$. Let $E \subset \Omega$ be closed in Ω and let $\mathcal{H}^{p+q-1}(E) = 0$. Let $w : \Omega \setminus E \to \mathbb{R}$ be separately subharmonic, that is,*

for all $y \in \mathbb{R}^q$ the function $(\Omega \setminus E)(y) \ni x \mapsto w(x,y) \in \mathbb{R}$ is subharmonic,

and

for all $x \in \mathbb{R}^p$ the function $(\Omega \setminus E)(x) \ni y \mapsto w(x,y) \in \mathbb{R}$ is subharmonic,

and such that the following conditions are satisfied:

(i) *$w \in \mathcal{L}^1_{\mathrm{loc}}(\Omega)$,*
(ii) *$w \in \mathcal{C}^2(\Omega \setminus E)$,*
(iii) *for each j, $1 \leq j \leq p$, $\frac{\partial^2 w}{\partial x_j^2} \in \mathcal{L}^1_{\mathrm{loc}}(\Omega)$ and for each k, $1 \leq k \leq q$, $\frac{\partial^2 w}{\partial y_k^2} \in \mathcal{L}^1_{\mathrm{loc}}(\Omega)$.*

Then w has a separately subharmonic extension to Ω.

Proof. Follows directly from Theorem 9.7 and from the above Lemma of Federer. □

9.5. **An extension result for harmonic functions.** For removability results for harmonic functions see, among others, [44, 45, 87, 150] and the references therein, say.

Now, using our Theorem 9.4, we give the following extension result for harmonic functions, compare with our previous version [121], Theorem 2, p. 155:

Theorem 9.9. *Suppose that Ω is a domain in $\mathbb{R}^n$, $n \geq 2$. Let $E \subset \Omega$ be closed in Ω and let $\mathcal{H}^{n-1}(E) < +\infty$. Let $u : \Omega \setminus E \to \mathbb{R}$ be harmonic and such that the following conditions are satisfied:*

(i) $u \in \mathcal{L}^1_{\mathrm{loc}}(\Omega)$.

(ii) *For each j, $1 \leq j \leq n$, $\frac{\partial^2 u}{\partial x_j^2} \in \mathcal{L}^1_{\mathrm{loc}}(\Omega)$.*

(iii) *For each j, $1 \leq j \leq n$, and for $\mathcal{H}^{n-1}$-almost all $X_j \in \mathbb{R}^{n-1}$ such that $E(X_j)$ is finite, the following condition holds:*
For each $x_j^0 \in E(X_j)$ there exist sequences $x_{j,l}^{0,1}, x_{j,l}^{0,2} \in (\Omega \setminus E)(X_j)$, $l = 1, 2, \ldots$, such that $x_{j,l}^{0,1} \nearrow x_j^0$, $x_{j,l}^{0,2} \searrow x_j^0$ as $l \to +\infty$, and

(iii(a)) $\lim_{l \to +\infty} u(x_{j,l}^{0,1}, X_j) = \lim_{l \to +\infty} u(x_{j,1}^{0,2}, X_j) \in \mathbb{R}$,

(iii(b)) $-\infty < \liminf_{l \to +\infty} \frac{\partial u}{\partial x_j}(x_{j,l}^{0,1}, X_j) = \limsup_{l \to +\infty} \frac{\partial u}{\partial x_j}(x_{j,l}^{0,2}, X_j) < +\infty$.

Then u has a unique harmonic extension to Ω.

Proof. Since the assumptions of Theorem 9.4 do hold for the subharmonic function u, u has a subharmonic extension u^* to Ω. On the other hand, the assumptions of Theorem 9.4 hold also for the subharmonic function $v = -u$. Thus $v = -u$ has a subharmonic extension $v^* = (-u)^*$ to Ω. As above in the proof of Theorem 9.4, we may suppose that the limits

$$\lim_{l \to +\infty} \left[\frac{\partial u}{\partial x_j}(x_{j,l}^{0,1}, X_j) \right] \quad \text{and} \quad \lim_{l \to +\infty} \left[\frac{\partial u}{\partial x_j}(x_{j,l}^{0,2}, X_j) \right]$$

indeed exist.

Since $-v^* = u^*$, the extension u^* of u is both subharmonic and superharmonic, thus harmonic and the claim follows. □

Example 21. The function $u : \mathbb{R}^2 \to \mathbb{R}$ given already in Example 19, shows that one cannot drop the condition (iii(b)) in Theorem 9.9.

9.5.1. Then a concise special case to Theorem 9.9:

Corollary 9.10. ([121], Corollary 3, p. 155) *Suppose that Ω is a domain in $\mathbb{R}^n$, $n \geq 2$. Let $E \subset \Omega$ be closed in Ω and let $\mathcal{H}^{n-1}(E) = 0$. Let $u : \Omega \setminus E \to \mathbb{R}$ be harmonic and such that the following conditions are satisfied:*

(i) $u \in \mathcal{L}^1_{\mathrm{loc}}(\Omega)$,

(ii) *for each j, $1 \leq j \leq n$, $\frac{\partial^2 u}{\partial x_j^2} \in \mathcal{L}^1_{\mathrm{loc}}(\Omega)$.*

Then u has a unique harmonic extension to Ω.

Proof. With the aid of the above Lemma one sees easily that the assumptions of Theorem 9.9 are satisfied. □

9.6. **An extension result for separately harmonic functions.** Next we give an extension result for separately harmonic functions. Our proof will be based on Theorem 9.7 and on basic properties of subharmonic functions.

Theorem 9.11. *Suppose that Ω is a domain in $\mathbb{R}^{p+q}$, $p,q \geq 2$. Let $E \subset \Omega$ be closed in Ω and let $\mathcal{H}^{p+q-1}(E) < +\infty$. Let $w : \Omega \setminus E \to \mathbb{R}$ be separately harmonic, that is,*

for all $y \in \mathbb{R}^q$ the function $(\Omega \setminus E)(y) \ni x \mapsto w(x,y) \in \mathbb{R}$ is harmonic,

and

for all $x \in \mathbb{R}^p$ the function $(\Omega \setminus E)(x) \ni y \mapsto w(x,y) \in \mathbb{R}$ is harmonic,

and such that the following conditions are satisfied:

(i) $w \in \mathcal{L}^1_{\text{loc}}(\Omega)$.

(ii) *For each j, $1 \leq j \leq p$, $\frac{\partial^2 w}{\partial x_j^2} \in \mathcal{L}^1_{\text{loc}}(\Omega)$, and for each k, $1 \leq k \leq q$, $\frac{\partial^2 w}{\partial y_k^2} \in \mathcal{L}^1_{\text{loc}}(\Omega)$.*

(iii) *For each j, $1 \leq j \leq p$, and for $\mathcal{H}^{p-1+q}$-almost all $(X_j, y) \in \mathbb{R}^{p-1+q}$ such that $E(X_j,y)$ is finite, the following condition holds:*
For each $x_j^0 \in E(X_j,y)$ there exist sequences $x_{j,l}^{0,1}, x_{j,l}^{0,2} \in (\Omega \setminus E)(X_j,y)$, $l = 1,2,\ldots$, such that $x_{j,l}^{0,1} \nearrow x_j^0$, $x_{j,l}^{0,2} \searrow x_j^0$ as $l \to +\infty$, and

(iii(a)) $\lim_{l\to+\infty} w(x_{j,l}^{0,1}, X_j, y) = \lim_{l\to+\infty} w(x_{j,l}^{0,2}, X_j, y) \in \mathbb{R}$,

(iii(b)) $-\infty < \liminf_{l\to+\infty} \frac{\partial w}{\partial x_j}(x_{j,l}^{0,1}, X_j, y) = \limsup_{l\to+\infty} \frac{\partial w}{\partial x_j}(x_{j,l}^{0,2}, X_j, y) < +\infty$.

(iv) *For each k, $1 \leq k \leq q$, and for $\mathcal{H}^{p+q-1}$-almost all $(x, Y_k) \in \mathbb{R}^{p+q-1}$ such that $E(x,Y_k)$ is finite, the following condition holds:*
For each $y_k^0 \in E(x,Y_k)$ there exist sequences $y_{k,l}^{0,1}, y_{k,l}^{0,2} \in (\Omega \setminus E)(x,Y_k)$, $l = 1,2,\ldots$, such that $y_{k,l}^{0,1} \nearrow y_k^0$, $y_{k,l}^{0,2} \searrow y_k^0$ as $l \to +\infty$, and

(iv(a)) $\lim_{l\to+\infty} w(x, y_{k,l}^{0,1}, Y_k) = \lim_{l\to+\infty} w(x, y_{k,l}^{0,2}, Y_k) \in \mathbb{R}$,

(iv(b)) $-\infty < \liminf_{l\to+\infty} \frac{\partial w}{\partial y_k}(x, y_{k,l}^{0,1}, Y_k) = \limsup_{l\to+\infty} \frac{\partial w}{\partial y_k}(x, y_{k,l}^{0,2}, Y_k) < +\infty$.

Then w has a separately harmonic extension to Ω.

Proof. By Lelong's theorem, see [66] or [49], Theorem, p. 54, separately harmonic functions are harmonic. Thus we know that $w \in \mathcal{C}^2(\Omega \setminus E)$. From Theorem 9.7 it then follows that w has a separately subharmonic extension $w^* : \Omega \to [-\infty, +\infty)$. Similarly, $u = -w$ has a separately subharmonic extension $u^* : \Omega \to [-\infty, +\infty)$. It is easy to see that $w^* = -u^*$. Thus w^* is in fact separately harmonic. □

Example 22. The function $u : \mathbb{R}^4 \to \mathbb{R}$ given already in Example 20, shows that one cannot drop the conditions (iii(b)) and (iv(b)) in Theorem 9.11.

Corollary 9.12. *Suppose that Ω is a domain in $\mathbb{R}^{p+q}$, $p,q \geq 2$. Let $E \subset \Omega$ be closed in Ω and let $\mathcal{H}^{p+q-1}(E) = 0$. Let $w : \Omega \setminus E \to \mathbb{R}$ be separately harmonic, that is,*

$$\text{for all } y \in \mathbb{R}^q \text{ the function } (\Omega \setminus E)(y) \ni x \mapsto w(x,y) \in \mathbb{R} \text{ is harmonic,}$$

and

$$\text{for all } x \in \mathbb{R}^p \text{ the function } (\Omega \setminus E)(x) \ni y \mapsto w(x,y) \in \mathbb{R} \text{ is harmonic,}$$

and such that the following conditions are satisfied:

(i) $w \in \mathcal{L}^1_{\text{loc}}(\Omega)$,

(ii) *for each* j, $1 \leq j \leq p$, $\dfrac{\partial^2 w}{\partial x_j^2} \in \mathcal{L}^1_{\text{loc}}(\Omega)$ *and for each* k, $1 \leq k \leq q$, $\dfrac{\partial^2 w}{\partial y_k^2} \in \mathcal{L}^1_{\text{loc}}(\Omega)$.

Then w *has a separately harmonic extension to* Ω.

Proof. Follows directly from Theorem 9.11 and from the above Lemma of Federer. □

10. EXTENSION RESULTS FOR PLURISUBHARMONIC AND FOR CONVEX FUNCTIONS

Abstract. We give extension results for plurisubharmonic, convex and separately convex functions. Our results improve previous results of Blanchet.

Keywords. Plurisubharmonic, convex, separately convex, Hausdorff measure

10.1. Previous results.

Blanchet's results [9], Theorems 3.1, 3.2 and 3.3, pp. 312–313, contain the following removability results for plurisubharmonic functions and for convex functions, see Theorem 9.2 above:

Theorem 10.1. *Let Ω be a domain in $\mathbb{R}^n$, $n \geq 2$, and let S be a hypersurface of class $\mathcal{C}^1$ which divides Ω into two subdomains Ω_1 and Ω_2. Let $u \in \mathcal{C}^0(\Omega) \cap \mathcal{C}^2(\Omega_1 \cup \Omega_2)$ be convex (or respectively plurisubharmonic provided Ω is then a domain in $\mathbb{C}^n$, $n \geq 1$) in Ω_1 and Ω_2. If $u_i = u|\Omega_i \in \mathcal{C}^1(\Omega_i \cup S)$, $i = 1, 2$, and*

$$\frac{\partial u_i}{\partial \overline{n}^k} \geq \frac{\partial u_k}{\partial \overline{n}^k} \tag{10.1}$$

on S with $i, k = 1, 2$, then u is convex (or respectively plurisubharmonic) in Ω.

We recall that above $\overline{n}^k = (\overline{n}_1^k, \overline{n}_2^k, \ldots, \overline{n}_n^k)$ is the unit normal exterior to Ω_k, and $u_k \in \mathcal{C}^1(\Omega_k \cup S)$, $k = 1, 2$, means that there exist n functions v_k^j, $j = 1, 2, \ldots, n$, continuous on $\Omega_k \cup S$, such that

$$v_k^j(x) = \frac{\partial u_k}{\partial x_j}(x)$$

for all $x \in \Omega_k$, $k = 1, 2$ and $j = 1, 2, \ldots, n$ (or respectively $j = 1, 2, \ldots, 2n$).

Moreover, and as shown already above in Example 18, the condition (10.1) in Blanchet's theorem cannot be dropped.

10.2. Improvements.

Now we improve our previous improvements still further, see Theorem 10.3, Theorem 10.6 and Corollary 10.8 below. Instead of hypersurfaces of class $\mathcal{C}^1$, we will below allow arbitrary sets of finite $(2n-1)$-dimensional or, respectively, of $(n-1)$-dimensional Hausdorff measure as exceptional sets. Then we must, however, replace the condition (10.1) by another, related condition, namely, in the case of plurisubharmonic functions by an additional integrability condition on the second partial derivatives $\frac{\partial^2 u}{\partial x_i \partial x_j}$, $i, j = 1, 2, \ldots, 2n$, or, respectively, in the case of convex functions, by a certain growth condition of left and right partial derivatives. Our methods of proof are rather elementary, thus natural, with the only exception that again we need Federer's geometric measure theory result, Lemma 9.1.

Juhani Riihentaus

10.3. **The case of plurisubharmonic functions.** In the proof of our result we use the following result of Lelong.

Lemma 10.2. ([67], Theorem 1, p. 18) *Suppose that* D *is a domain of* $\mathbb{C}^n$, $n \geq 2$. *Let* $v : D \to [-\infty, +\infty)$. *Then* v *is plurisubharmonic if and only if the following condition holds:*
For each $z_0 \in D$ *and for each affine transformation* $A = (A_1, A_2, \dots, A_n) : \mathbb{C}^n \to \mathbb{C}^n$,

$$z' = Az \Leftrightarrow (z'_1, z'_2, \dots, z'_n) = (A_1(z_1, z_2, \dots, z_n), A_2(z_1, z_2, \dots, z_n), \dots, A_n(z_1, z_2, \dots, z_n))$$

$$\Leftrightarrow \begin{cases} z'_1 = A_1(z_1, z_2, \dots, z_n) = z_1^0 + a_{11}z_1 + a_{12}z_2 + \cdots + a_{1n}z_n, \\ z'_2 = A_2(z_1, z_2, \dots, z_n) = z_2^0 + a_{21}z_1 + a_{22}z_2 + \cdots + a_{2n}z_n, \\ \vdots \\ z'_n = A_n(z_1, z_2, \dots, z_n) = z_n^0 + a_{n1}z_1 + a_{n2}z_2 + \cdots + a_{nn}z_n, \end{cases}$$

for which $\det A \neq 0$, *the function* $v \circ A^{-1} : A(D) \to [-\infty, +\infty)$ *is subharmonic.*

Then the result:

Theorem 10.3. *Suppose that* Ω *is a domain of* $\mathbb{C}^n$, $n \geq 2$. *Let* $E \subset \Omega$ *be closed in* Ω *and* $\mathcal{H}^{2n-1}(E) < +\infty$. *Let* $u : \Omega \to \mathbb{R}$ *be such that*

(i) $u \in \mathcal{C}^1(\Omega)$,
(ii) $u \in \mathcal{C}^2(\Omega \setminus E)$,
(iii) *for each* i, j, $1 \leq i, j \leq 2n$, $\dfrac{\partial^2 u}{\partial x_i \partial x_j} \in \mathcal{L}^1_{\mathrm{loc}}(\Omega)$,
(iv) u *is plurisubharmonic in* $\Omega \setminus E$.

Then u *is plurisubharmonic.*

Proof. By Lemma 10.2 it is sufficient to show that $v = u \circ A$ is subharmonic in $\Omega' = A^{-1}(\Omega)$ for any affine mapping $A : \mathbb{C}^n \to \mathbb{C}^n$ with $\det A \neq 0$. Clearly $v \in \mathcal{C}^1(\Omega')$ and $v \in \mathcal{C}^2(\Omega' \setminus E')$, where $E' = A^{-1}(E)$. It is easy to see that for each j, $1 \leq j \leq 2n$, $\dfrac{\partial^2 v}{\partial x_j^2} \in \mathcal{L}^1_{\mathrm{loc}}(\Omega')$. Since u is plurisubharmonic in $\Omega \setminus E$, v is by Lemma 10.2 subharmonic in $\Omega' \setminus E'$, thus subharmonic in Ω' by Corollary 9.5. □

10.4. **The case of convex functions.** We recall here some very basic properties of convex functions. Let D be a domain of $\mathbb{R}^n$, $n \geq 1$. A function $f : D \to \mathbb{R}$ is *convex* if the following condition is satisfied: For each $x, y \in D$ such that $\{tx + (1-t)y : t \in [0,1]\} \subset D$, one has $f(tx + (1-t)y) \leq t\, f(x) + (1-t)\, f(y)$ for all $t \in [0,1]$.

Lemma 10.4. ([154], Theorem 5.1.3, p. 195) *Let* $f : [a,b] \to \mathbb{R}$ *be a convex function. Then* f *possesses left and right derivatives at each interior point of* $[a,b]$, *and if* x_1, x_2 *are interior points of* $[a,b]$ *with* $x_1 < x_2$, *then*

$$-\infty < f'_-(x_1) \leq f'_+(x_1) \leq \frac{f(x_2) - f(x_1)}{x_2 - x_1} \leq f'_-(x_2) \leq f'_+(x_2) < +\infty.$$

Lemma 10.5. ([154], Theorem 5.1.8, p. 198) *Let $f:(a,b)\to\mathbb{R}$. Then f is convex if and only if it has support at each point of (a,b), i.e. for any $x_0\in(a,b)$ there is a constant $m\in\mathbb{R}$ such that*

$$f(x_0)+m(x-x_0)\leq f(x)$$

for all $x\in(a,b)$.

Moreover, if f is convex, then any m, $f'_-(x_0)\leq m\leq f'_+(x_0)$, will do.

We consider first separately convex functions:

Theorem 10.6. *Suppose that Ω is a domain of $\mathbb{R}^n$, $n\geq 2$. Let $E\subset\Omega$ be closed in Ω and $\mathcal{H}^{n-1}(E)<+\infty$. Let $u:\Omega\to\mathbb{R}$ be such that*

(i) *$u\in\mathcal{C}^0(\Omega)$,*
(ii) *for each j, $1\leq j\leq n$, and for $\mathcal{H}^{n-1}$-almost all $X_j\in\mathbb{R}^{n-1}$ such that $E(X_j)$ is finite, one has*

$$\liminf_{\varepsilon\to 0+0}\frac{\partial_- u}{\partial x_j}(x_j^0-\varepsilon,X_j)\leq\limsup_{\varepsilon\to 0+0}\frac{\partial_+ u}{\partial x_j}(x_j^0+\varepsilon,X_j)$$

for each $x_j^0\in E(X_j)$,
(iii) *u is separately convex in $\Omega\setminus E$.*

Then u is separately convex.

Above, and in the sequel, $\frac{\partial_- u}{\partial x_j}(x_j,X_j)$ and $\frac{\partial_+ u}{\partial x_j}(x_j,X_j)$ are the left and right partial derivatives of u, respectively, taken at the point $x=(x_j,X_j)$.

Remark 10.7. By Lemma 10.4 the condition (ii) is a necessary condition for (separately) convex functions.

Proof. To prove Theorem 10.6 choose j, $1\leq j\leq n$, arbitrarily. Using Lemma 9.1 and the condition (ii) we see that for $\mathcal{H}^{n-1}$-almost all $X_j\in\mathbb{R}^{n-1}$,

$$\begin{cases} E(X_j) \ \text{ is finite,} \\ (\Omega\setminus E)(X_j)\ni x_j\mapsto u(x_j,X_j)\in\mathbb{R} \ \text{ is convex,} \\ \liminf\limits_{\varepsilon\to 0+0}\frac{\partial_- u}{\partial x_j}(x_j^0-\varepsilon,X_j)\leq\limsup\limits_{\varepsilon\to 0+0}\frac{\partial_+ u}{\partial x_j}(x_j^0+\varepsilon,X_j) \ \text{ for all } x_j^0\in E(X_j). \end{cases}$$

Take $X_j\in\mathbb{R}^{n-1}$ and write $E(X_j)=\{x_1^o,x_2^o,\ldots,x_N^o\}$. Let (a,b) be an arbitrary component of $\Omega(X_j)$. It is sufficient to show that $u(\cdot,X_j)|(a,b)$ is convex.

Divide the interval (a,b) into parts (a_k,b_k), $k=1,2,\ldots,N$ such that

$$x_j^o\in(a_j,b_j) \text{ and } b_j=a_{j+1},\ j=1,2,\ldots,N-1,\ a_1=a,\ b=b_N.$$

We show that $u|(a_k,b_k)$ is convex. Write $a=a_k$, $b=b_k$ and $u=u(\cdot,X_j)|(a_1,b_N)$. We know that $u_1=u|(a,x_o)$ and $u_2=u|(x_o,b)$ are convex. We begin by showing that u_1 has a support at each point of (a,b), that is, for each $t_o\in(a,x_o)$,

$$u_t^-(x):=u(t)+u'_-(t)(x-t)\leq u(x) \text{ for each } x,\ a<x<b.$$

We know that

$$u_t^-(x) := u(t) + u'_-(t)(x-t) \leq u(x) \text{ for each } x,\ a < x < x_o.$$

It remains to show that

$$u_t^-(x) := u(t) + u'_-(t)(x-t) \leq u(x) \text{ for each } x,\ x_o \leq x < b.$$

Below we use the following notation (remembering also Lemma 10.4):

$$m_1 := \liminf_{\varepsilon \to 0+0} u'_-(x_o - \varepsilon) = \lim_{\varepsilon \to 0+0} u'_-(x_o - \varepsilon) \in \mathbb{R},$$

$$m_2 := \limsup_{\varepsilon \to 0+0} u'_-(x_o + \varepsilon) = \lim_{\varepsilon \to 0+0} u'_-(x_o + \varepsilon) \in \mathbb{R},$$

$$u_t^-(x) := u(t) + u'_-(t)(x-t),\ u_t^+(x) := u(t) + u'_+(t)(x-t),$$

$$u_{x_o}^-(x) := u(x_o) + m_1(x - x_o),\ u_t^+(x) := u(t) + m_2(x - x_o)$$

It is easy to see that $u_t^-(t) = u(t)$ and $u_t^-(x_o) \leq u(x_o)$. Similarly, $u_{x_o}^-(t) \leq u(t)$ and $u_{x_o}^-(x_o) \leq u(x_o)$.

Next we compare the functions u_t^- and $u_{x_o}^-$ when $a < t < x_o$. Since $u_t^-(t) = u(t) \geq u_{x_o}^-(t)$ and $u_t^-(x_o) \leq u(x_o) = u_{x_o}^-(x_o) \leq u(x_o)$, there is x_3, $t \leq x_3 \leq x_o$, such that

$$u_t^-(x) \leq u_{x_o}^-(x) \leq u_{x_o}^+(x) \leq u(x) \ \text{ for all } \ x > x_3.$$

Hence we get the demanded inequality,

$$u_t^-(x) := u(t) + u'_-(t)(x-t) \leq u(x) \ \text{ for each } \ x,\ x_o \leq x < b.$$

Next we show that u_2 has a support at each point of (a,b), that is, for each $t \in (x_0, b)$,

$$u_t^+(x) := u(t) + u'_+(t)(x-t) \leq u(x) \ \text{ for each } \ x,\ a < x < b.$$

We must show that

$$u_t^-(x) := u(t) + u'_-(t)(x-t) \leq u(x) \ \text{ for } \ \text{ each } x,\ a < x < x_0.$$

It is sufficient to show that

$$u_t^+(x) := u(t) + u'_+(t)(x-t) \leq u(x) \ \text{ for each } \ x,\ a < x \leq x_0.$$

We proceed as above. We know that

$$u_t^+(x) := u(t) + u'_+(t)(x-t) \leq u(x) \ \text{ for each } \ x,\ x_0 < x < b.$$

Suppose that $x_0 \leq x \leq b$ and let $t \searrow x_0$. Then we get

$$u_{x_0}^+(x) = u(x_0) + m_2(x - x_0) \leq u(x) \ \text{ for each } \ x,\ a < x \leq x_0.$$

As a matter of fact, we have

$$u_{x_0}^-(x) = u(x_0) + m_1(x - x_0) \leq u_{x_0}^+(x) = u(x_o) + m_2(x - x_0) \leq u(x)$$

for each x, $x_0 < x < b$. Thus we have $u_t^+(t) = u(t) \geq u_{x_0}^+(t)$. Also $u_t^+(x_0) \leq u(x_0)$. Therefore, taking the limit $t \searrow x_0$, we get

$$u_{x_0}^+(x) = u(x_0) + m_2(x - x_0) \leq u(x) \ \text{ for each } \ x,\ a < x \leq x_0.$$

Taking once more the limit $x \searrow x_0$ (or just a direct substitution!), $u_{x_0}^+(x_0) = u(x_0)$. Thus we have

$$\begin{cases} u_t^+(t) = u(t) \geq u_{x_0}^+(t), \\ u_t^+(x_0) \leq u(x_0) = u_{x_0}(x_0). \end{cases}$$

Therefore there is x_4, $x_0 < x_4 < t$, such that $u_t^+(x) \leq u_{x_0}^+(x)$ whenever $x \leq x_4$. But then we have

$$u_t^+(x) \leq u_{x_0}^+(x) \leq u_{x_0}^-(x) \leq u_{x_0}^+(x) \leq u(x),$$

because we have

$$u_{x_0}^+(x) = u(x_0) + m_2(x - x_0) \leq u(x_0) + m_1(x - x_0) = u_{x_0}^-(x).$$

Thus the claim is proved and we know that $u|(a_k, b_k)$ is convex.

Next we show that if $u|(a_1, b_1)$ and $u|(a_2, b_2)$ are convex, then $u|(a_1, b_2)$ is convex. We know that

$$a_1 < t_1 < b_1 \Rightarrow u_{t_1}(x) := u(t_1) + u'_-(t_1)(x - t_1) \leq u(x) \text{ for all } x,\ a_1 < x < b_1,$$

and

$$a_2 < t_2 < b_2 \Rightarrow u_{t_2}(x) := u(t_2) + u'_-(t_2)(x - t_2) \leq u(x) \text{ for all } x,\ a_2 < x < b_2.$$

As a matter of fact, instead of $u'_-(t_1)$ one can use any real number A_1 such that

$$u'_-(t_1) \leq A_1 \leq u'_+(t_1).$$

Similarly, instead of $u'_-(t_2)$ one can use any real number A_2 such that

$$u'_-(t_2) \leq A_2 \leq u'_+(t_2).$$

We must show that

$$a_1 < t_1 < b_1 \Rightarrow u_{t_1}(x) := u(t_1) + u'_-(t_1)(x - t_1) \leq u(x) \text{ for all } x,\ b_1 \leq x < b_2,$$

and

$$a_2 < t_2 < b_2 \Rightarrow u_{t_2}(x) := u(t_2) + u'_-(t_2)(x - t_2) \leq u(x) \text{ for all } x,\ a_1 < x \leq b_1.$$

First the "first case"! Begin with the case $x = b_1$. We know that

$$u_{t_1}(x) := u(t_1) + u'_-(t_1)(x - t_1) \leq u(x) \text{ for all } x,\ a_1 \leq x < b_1.$$

Taking the limit $x \nearrow b_1$ we get

$$u_{t_1}(b_1) := u(t_1) + u'_-(t_1)(b_1 - t_1) \leq u(b_1)$$

Then the case $b_1 < x < b_2$. Here we must show that

$$u_{t_1}^-(x) := u(t_1) + u'_-(t_1)(x - t_1) \leq u(x) \text{ for all } x,\ b_1 = a_2 \leq x < b_2.$$

One sees easily that

$$u_{t_1}^-(t_1) = u(t_1) \text{ and } u_{t_1}^-(b_1) \leq u(b_1).$$

Next we consider the situation at the point $b_1 = a_2$. We know that

$$u_{t_1}^-(x) := u(t_1) + u'_-(t_1)(x - t_1) \leq u(x) \text{ for all } x,\ a_1 \leq x < b_1.$$

Taking here the limit $t_1 \nearrow b_1$,

$$u_{b_1}^-(x) := u(b_1) + \lim_{t_1 \to b_1} u'_-(t_1)(x - b_1) \leq u(x) \text{ for all } x,\ a_1 \leq x < b_1$$

or

$$u_{b_1}^-(x) := u(b_1) + m_1(x - b_1) \leq u(x) \text{ for all } x,\ a_1 \leq x < b_1$$

Thus $u_{b_1}^-(b_1) = u(b_1)$ and $u_{b_1}^-(t_1) \leq u(t_1)$. Therefore, there exists x_3, $t_1 < x_3 < b_1$, such that

$$x > x_3 \Rightarrow u_{t_1}(x) := u(t_1) + u'_-(t_1)(x - t_1) \leq$$
$$\leq u_{b_1}^-(x) := u(b_1) + m_1(x - b_1) \leq u(b_1) + m_2(x - b_1) \text{ for all } x, x > b_1.$$

Therefore the following:

$$u_{t_1}(x) = u(t_1) + u'_-(t_1)(x - t_1) \leq u_{b_1}^-(x) :=$$
$$:= u(b_1) + m_1(x - b_1) \leq u(b_1) + m_2(x - b_1) =: u_{b_1}^+(x) \leq u(x) \text{ for all } x, x > b_1.$$

This is seen as follows: We know that

$$u_{t_2}^+(x) := u(t_2) + u'_+(t_2)(x - t_2) \leq u(x) \text{ for all } x,\ a_2 < x \leq b_2$$

Taking the limit $t_2 \searrow b_1$, we get

$$u_{b_1}^+(x) := u(b_1) + m_2(x - b_1) \leq u(x) \text{ for all } x,\ a_2 < x \leq b_2.$$

Thus we have obtained that

$$u_{b_1}^+(x) \leq u(x) \text{ for all } x,\ a_2 = b_1 < x \leq b_2.$$

Next we consider the base at the point t_2, $b_1 = a_2 < t_2 < b_2$. We must show that

$$u_{t_2}^-(x) := u(t_2) + u'_-(t_2)(x - t_2) \leq u(x) \text{ for all } x,\ a_1 < x \leq b_2$$

Using this, Lemma 10.4 and Lemma 10.5 one sees that for $\mathcal{H}^{n-1}$-almost all $X_j \in \mathbb{R}^{n-1}$ the functions

$$\Omega(X_j) \ni x_j \mapsto u(x_j, X_j) \in \mathbb{R} \tag{10.2}$$

are in fact convex. From this and from the fact that u is continuous, it follows that the functions of the form (10.2) are convex *for all* $X_j \in \mathbb{R}^{n-1}$. Since j, $1 \leq j \leq n$, was arbitrary the claim follows. □

Corollary 10.8. *Suppose that Ω is a domain of $\mathbb{R}^n$, $n \geq 2$. Let $E \subset \Omega$ be closed in Ω and $\mathcal{H}^{n-1}(E) < +\infty$. Let $u : \Omega \to \mathbb{R}$ be such that*

(i) $u \in \mathcal{C}^1(\Omega)$,
(ii) *u is (separately) convex in $\Omega \setminus E$.*

Then u is (separately) convex.

Proof. The separately convex case follows directly from Theorem 10.6. The convex case follows from the separately convex case with the aid of the following Lelong type result (whose proof is similar to [67], proof of Theorem 1, p. 18). □

Lemma 10.9. *Suppose that D is a domain of $\mathbb{R}^n$, $n \geq 2$. Let $v : D \to [-\infty, +\infty)$. Then v is convex if and only if the following condition holds:*
For each $x_0 \in D$ and for each affine transformation $A = (A_1, A_2, \ldots, A_n) : \mathbb{R}^n \to \mathbb{R}^n$,

$$x' = Ax \Leftrightarrow (x'_1, x'_2, \ldots, x'_n) = (A_1(x_1, x_2, \ldots, x_n), A_2(x_1, x_2, \ldots, x_n), \ldots, A_n(x_1, x_2, \ldots, x_n))$$

$$\Leftrightarrow \begin{cases} x'_1 = A_1(x_1, x_2, \ldots, x_n) = x_1^0 + a_{11}x_1 + a_{12}x_2 + \cdots + a_{1n}x_n, \\ x'_2 = A_2(x_1, x_2, \ldots, x_n) = x_2^0 + a_{21}x_1 + a_{22}x_2 + \cdots + a_{2n}x_n, \\ \vdots \\ x'_n = A_n(x_1, x_2, \ldots, x_n) = x_n^0 + a_{n1}x_1 + a_{n2}x_2 + \cdots + a_{nn}x_n, \end{cases}$$

for which $\det A \neq 0$, the function $v \circ A^{-1} : A^{-1}(D) \to [-\infty, +\infty)$ is separately convex.

11. Extension Results for Holomorphic and for Meromorphic Functions

Abstract. Applying our extension result for subharmonic functions, we give an extension result for holomorphic functions, which is related to the well-known extension results of Besicovitch and Shiffman. In addition, we give related older extension results of holomorphic functions and of meromorphic functions.

Keywords. Holomorphic, meromorphic, Hausdorff measure

11.1. Extension results for holomorphic functions with the aid of Theorem 9.4.

Our result is the following, see also our preliminary result [121], Theorem 3, p. 156.

Theorem 11.1. *Suppose that Ω is a domain in $\mathbb{C}^n$, $n \geq 1$. Let $E \subset \Omega$ be closed in Ω and let $\mathcal{H}^{2n-1}(E) < +\infty$. Let $f = u + iv : \Omega \setminus E \to \mathbb{C}$ be holomorphic and such that the following conditions are satisfied:*

(i) $f \in \mathcal{L}^1_{\text{loc}}(\Omega)$.

(ii) *For each j, $1 \leq j \leq 2n$, $\frac{\partial^2 u}{\partial x_j^2} \in \mathcal{L}^1_{\text{loc}}(\Omega)$ and $\frac{\partial^2 v}{\partial x_j^2} \in \mathcal{L}^1_{\text{loc}}(\Omega)$.*

(iii) *For each j, $1 \leq j \leq 2n$, and for $\mathcal{H}^{2n-1}$-almost all $X_j \in \mathbb{R}^{2n-1}$ such that $E(X_j)$ is finite, the following condition holds:*
For each $x_j^0 \in E(X_j)$ there exist sequences $x_{j,l}^{0,1}, x_{j,l}^{0,2} \in (\Omega \setminus E)(X_j)$, $l = 1, 2, \ldots$, such that $x_{j,l}^{0,1} \nearrow x_j^0$, $x_{j,l}^{0,2} \searrow x_j^0$ as $l \to +\infty$, and

(iii(a)) $\lim_{l \to +\infty} f(x_{j,l}^{0,1}, X_j) = \lim_{l \to +\infty} f(x_{j,l}^{0,2}, X_j) \in \mathbb{C}$,

(iii(b)) $-\infty < \liminf_{l \to +\infty} \frac{\partial u}{\partial x_j}(x_{j,l}^{0,1}, X_j) = \limsup_{l \to +\infty} \frac{\partial u}{\partial x_j}(x_{j,l}^{0,2}, X_j) < +\infty$ *and*
$-\infty < \liminf_{l \to +\infty} \frac{\partial v}{\partial x_j}(x_{j,l}^{0,1}, X_j) = \limsup_{l \to +\infty} \frac{\partial v}{\partial x_j}(x_{j,l}^{0,2}, X_j) < +\infty$.

Then f has a unique holomorphic extension to Ω.

Proof. It is sufficient to show that u and v have harmonic extensions u^* and v^* to Ω. As a matter of fact, then $f^* = u^* + iv^* : \Omega \to \mathbb{C}$ is in $\mathcal{C}^\infty(\Omega)$ and thus a continuous function. Therefore the claim follows from Shiffman's theorem or also from [89, 90].

Another possibility for the proof is just to observe that the in $\Omega \setminus E$ harmonic functions u and v have by Theorem 9.9 harmonic extensions u^* and v^* to Ω. Since u^* and v^* are thus $\mathcal{C}^\infty$ functions, the holomorphy of the extension $f^* = u^* + iv^*$ in Ω follows easily. □

11.1.1. As a concise corollary we get the following:

Corollary 11.2. ([120], Theorem 3, p. 51, [121], Theorem 4, p. 157) *Suppose that Ω is a domain in $\mathbb{C}^n$, $n \geq 1$. Let $E \subset \Omega$ be closed in Ω and let $\mathcal{H}^{2n-1}(E) = 0$. Let $f : \Omega \setminus E \to \mathbb{C}$ be holomorphic and such that the following conditions are satisfied:*

(i) $f \in \mathcal{L}^1_{\text{loc}}(\Omega)$,

(ii) *for each j, $1 \leq j \leq 2n$, $\frac{\partial^2 f}{\partial x_j^2} \in \mathcal{L}^1_{\text{loc}}(\Omega)$.*

Then f has a unique holomorphic extension to Ω.

11.1.2. Observe that our Theorem 11.1 and Corollary 11.2 can be considered, at least in some sense, as counterparts to two of Shiffman's well-known extension results for holomorphic functions, namely the following results:

Theorem 11.3. ([135], Lemma 3, p. 115, and [45], Theorem 1.1 (b), p. 703) *Let Ω be a domain in $\mathbb{C}^n$, $n \geq 1$. Let $E \subset \Omega$ be closed in Ω and let $\mathcal{H}^{2n-1}(E) < +\infty$. If $f : \Omega \to \mathbb{C}$ is continuous and $f|\Omega \setminus E$ is holomorphic, then f is holomorphic in Ω.*

Theorem 11.4. ([135], Lemma 3, p. 115, and [45], Theorem 1.1 (c), p. 703) *Let Ω be a domain in $\mathbb{C}^n$, $n \geq 1$. Let $E \subset \Omega$ be closed in Ω and let $\mathcal{H}^{2n-1}(E) = 0$. If $f : \Omega \setminus E \to \mathbb{C}$ is holomorphic and bounded, then f has a unique holomorphic extension to Ω.*

Shiffman's proofs of his above results were based on coordinate rotation, on the use of Cauchy integral formula, on the already stated important result of Federer, Lemma 9.1 above, and on the following classical result of Besicovitch:

Theorem 11.5. ([8], Theorem 1, p. 2) *Let D be a domain in $\mathbb{C}$. Let $E \subset D$ be closed in D and let $\mathcal{H}^1(E) = 0$. If $f : D \setminus E \to \mathbb{C}$ is holomorphic and bounded, then f has a unique holomorphic extension to D.*

For slightly more general versions of Shiffman's results with different proofs, see [89], Theorem 3.1, p. 49, Theorem 3.5, p. 52, Corollary 3.7, p. 54, and [90], Theorem 3.1, p. 333, Corollary 3.3, p. 336.

Next we give some related, previous results. However and for this purpose, we recall first certain types of exceptional sets.

11.2. **Exceptional sets.** Let δ be a non-negative set function defined on the subsets of $\mathbb{C}$ such that

(a) if $A \subset B$, then $\delta(A) \leq \delta(B)$;

(b) if K_j, $j = 1, 2, \ldots$, are compact subsets of $\mathbb{C}$, then
$$\delta\left(\cup_{j=1}^{+\infty} K_j\right) = 0$$
if and only if $\delta(K_j) = 0$ for each $j = 1, 2, \ldots$;

(c) if K_j, $j = 1, 2, \ldots$, is a decreasing sequence of compact subsets of $\mathbb{C}$, then
$$\delta\left(\cap_{j=1}^{+\infty} K_j\right) = \lim_{j \to +\infty} \delta(K_j).$$

To define *n-small* sets in $\mathbb{C}^n$, we use induction. For each set $F \subset \mathbb{C}$, set $\mathcal{D}^1_\delta(F) = \delta(F)$. If $n \geq 2$ and $\mathcal{D}^{n-1}_\delta$ is defined for subsets of $\mathbb{C}^{n-1}$, we set for $F \subset \mathbb{C}^n$,
$$\mathcal{D}^n_\delta(F) = \max_{1 \leq j \leq n} \mathcal{H}^2\{z_j \in \mathbb{C} : \mathcal{D}^{n-1}_\delta\{Z_j \in \mathbb{C}^{n-1} : (z_j, Z_j) \in F\} > 0\}.$$
We say that $F \subset \mathbb{C}^n$ is *n-small in* δ if $\mathcal{D}^n_\delta(F) = 0$.

In the sequel δ will be either the outer logarithmic capacity cap^* or the analytic capacity γ. If $\mathcal{D}^n_{cap^*}(F) = 0$ we say shortly that F is *n-small*.

Proposition 11.6. ([92], Definition 2.2, p. 101, and [50], 2.2, p. 472) *Let δ be as above. Then a set $F \subset \mathbb{C}^n$, $n \geq 2$, is n-small in δ, if for each j, $1 \leq j \leq n$, $\mathcal{H}^{2n-2}(F_j) = 0$, where*

$$F_j = \{Z_j \in \mathbb{C}^{n-1} : \delta\{z_j \in \mathbb{C} : (z_j, Z_j) \in F\} > 0\}.$$

Conversely, if F is n-small in δ and an $\mathcal{F}_\sigma$-set, then $\mathcal{H}^{2n-2}(F_j) = 0$ for each j, $1 \leq j \leq n$.

Remark 11.7. Clearly n-small sets are n-small in analytic capacity, but not conversely.

Remark 11.8. Sets of zero $(2n-1)$-dimensional Hausdorff measure are n-small in analytic capacity. This follows from the above Proposition 11.6.

Remark 11.9. Using a result of Mattila [70], Corollary 3.3, p. 263, see also [129], Lemma 6, p. 115, we see with the aid of Proposition 11.6 that polar sets (in $\mathbb{R}^{2n}$) are n-small. Thus also sets of locally finite $(2n-2)$-dimensional Hausdorff measure are n-small. See [92], Remark 2.6, p. 102.

Since $\mathcal{H}^2$ is an outer measure, we see using the subadditivity of the outer logarithmic capacity and induction that $\mathcal{D}^n_\delta$ is an outer measure in $\mathbb{C}^n$. Thus $F \subset \mathbb{C}^n$ is n-small if and only if $F \cap U$ is n-small for each $U \subset \mathbb{C}^n$ open. Similarly, using induction we see that $F \subset \mathbb{C}^n$ is of zero Lebesgue measurable if F is Lebesgue measurable and n-small. There are, however, n-small sets which are Lebesgue nonmeasurable, see [92], Remark 2.8, p. 103.

Remark 11.10. If $\delta = cap^*$ and if one, moreover, replaces above the Hausdorff measure $\mathcal{H}^2$ by cap^*, one gets the set function g_n, see [13], p. 284. Thus

$$g_n(F) = \max_{1 \leq j \leq n} \{cap^*\{z_j \in \mathbb{C} : g_{n-1}(\{Z_j \in \mathbb{C}^{n-1} : (z_j, Z_j) \in F\}) > 0\}\}.$$

Clearly, the sets F for which $g_n(F) = 0$, are n-small, but the converse does not hold.

However, not even sets E with $g_n(F) = 0$ need be polar. In fact, with the help of [20], Theorem 2, p. 118, it is possible to construct a compact set $F \subset \mathbb{C}^2$ of Hausdorff dimension 4 such that $g_n(F) = 0$ and for each $z \in \mathbb{C}$ the sets $F \cap (\mathbb{C} \times \{z\})$ and $F \cap (\{z\} \times \mathbb{C})$ contain at most one point.

11.3. **A previous, slightly related result.** Observe that, in addition to Corollary 11.2, also the following result holds:

Theorem 11.11. *Suppose that Ω is a domain in $\mathbb{C}^n$, $n \geq 1$. Let $F \subset \Omega$ be closed in Ω and let $\mathcal{H}^{2n-1}(F) = 0$. Let $f : \Omega \setminus F \to \mathbb{C}$ be holomorphic. If for each j, $1 \leq j \leq 2n$, $\frac{\partial f}{\partial x_j} \in \mathcal{L}^2_{\mathrm{loc}}(\Omega)$, then f has a holomorphic extension to Ω.*

The proof follows at once from the following, rather old result. Observe that, in addition to Corollary 11.13, also the following more general result holds:

Theorem 11.12. ([51], Theorem 3.5, pp. 300-301) *Suppose that Ω is a domain in $\mathbb{C}^n$, $n \geq 1$. Let $F \subset \Omega$ be closed in Ω and n-small in analytic capacity. Let $f : \Omega \setminus F \to \mathbb{C}$*

be holomorphic. If for some $p \in \mathbb{R}$,

$$\int_{\Omega \setminus F} |f(z)|^{p-2} \sum_{j=1}^{n} |\frac{\partial f}{\partial z_j}(z)|^2 \, dm_{2n}(z) < +\infty, \tag{11.1}$$

then f *has a meromorphic extension* f^* *to* Ω. *If* $p \geq 0$, *then* f^* *is holomorphic.*

Proof. Consider first the case $n = 1$. Then clearly

$$\int_{f(\Omega \setminus F)} |w|^{p-2} \, dm_2(w) \leq \int_{\Omega \setminus F} |f(z)|^{p-2} |f'(z)|^2 dm_2(z). \tag{11.2}$$

Since for each $p \in \mathbb{R}$,

$$\int_{\mathbb{C}} |w|^{p-2} dm_2(w) = +\infty, \tag{11.3}$$

it follows from these that f omits a set of positive measure. Thus by a result of Kametani, [76], Theorem 2, p. 10, and Remark, p. 11, f has a meromorphic extension f^* to Ω. If $p \geq 0$, (11.1) and (11.2) imply that f^* cannot have poles and is thus holomorphic.

Suppose then that $n \geq 2$. By Fubini's theorem there is for each j, $1 \leq j \leq n$, a set $B_j \subset \mathbb{C}^{n-1}$ such that $\mathcal{H}^{2n-2}(B_j) = 0$ and

$$\int_{(\Omega \setminus F)(Z_j)} |f(z_j, Z_j)|^{p-2} |\frac{\partial f}{\partial z_j}(z_j, Z_j)|^2 \, dm_2(z_j) < +\infty, \tag{11.4}$$

whenever $Z_j \notin B_j$. Since F is n-small in analytic capacity, we may, by Proposition 11.6 above, suppose that the section $F(Z_j)$ is of zero analytic capacity for each $Z_j \notin B_j$.

By the case $n = 1$ considered above, the holomorphic functions

$$f_{Z_j} : \Omega(Z_j) \setminus F(Z_j) \longrightarrow \mathbb{C}, \qquad Z_j \notin B_j,$$

have meromorphic extensions $f^*_{Z_j}$ to $\Omega(Z_j)$. Moreover, these extensions are holomorphic if $p \geq 0$. Thus the theorem follows from Lemma 11.15, see below. □

Corollary 11.13. ([51], Theorem 3.5, pp. 300-301) *Suppose that* Ω *is a domain in* $\mathbb{C}^n$, $n \geq 1$. *Let* $F \subset \Omega$ *be closed in* Ω *and let* $\mathcal{H}^{2n-1}(F) = 0$. *Let* $f : \Omega \setminus F \to \mathbb{C}$ *be holomorphic. If for some* $p \in \mathbb{R}$,

$$\int_{\Omega \setminus F} |f(z)|^{p-2} \sum_{j=1}^{n} |\frac{\partial f}{\partial z_j}(z)|^2 \, dm_{2n}(z) < +\infty, \tag{11.5}$$

then f *has a meromorphic extension* f^* *to* Ω. *If* $p \geq 0$, *then* f^* *is holomorphic.*

Remark 11.14. The holomorphic function $f : \mathbb{C}^{n-1} \times (B^2(0,1) \setminus \{0\} \to \mathbb{C}$,

$$f(z', z_n) = \frac{1}{z_n},$$

shows that condition (11.5) with $p < 0$ gives only a meromorphic (not holomorphic) extension.

Lemma 11.15. ([51], Lemma 3.4, pp. 299-300) *Let $n \geq 2$. Let $F \subset \Omega$ be such that for each j, $1 \leq j \leq n$, and for $\mathcal{H}^{2n-2}$-almost all $Z_j \in \mathbb{C}^{n-1}$ the section $F(Z_j)$ is totally disconnected. Let $f : \Omega \setminus F \to \mathbb{C}$ be a holomorphic function. If for each j, $1 \leq j \leq n$, and for $\mathcal{H}^{2n-2}$-almost all $Z_j \in \mathbb{C}^{n-1}$ the function $f_{Z_j} : (\Omega \setminus F)(Z_j) \longrightarrow \mathbb{C}$,*

$$f_{Z_j}(z_j) = f(z_j, Z_j),$$

has a holomorphic (respectively meromorphic) extension to $\Omega(Z_j)$, then f has a holomorphic (respectively meromorphic) extension to Ω.

For related, partly previous and partly more general results, see [13], Theorem, p. 284, [51], Theorem 3.5, pp. 300-301, and [95], Theorem 3.1, pp. 925-926.

11.4. **Additional extension results for holomorphic and for meromorphic functions.** We give below a partly extended version of the above result:

Theorem 11.16. ([95], Theorem 3.1, pp. 925-926) *Suppose that Ω is a domain in $\mathbb{C}^n$, $n \geq 1$. Let $F \subset \Omega$ be closed in Ω and let $\mathcal{H}^{2n-1}(F) = 0$. Let $f : \Omega \setminus F \to \mathbb{C}$ be holomorphic. If for some $p \in \mathbb{R}$,*

$$\int_{\Omega \setminus F} \frac{|f(z)|^{p-2}}{(1+|f(z)|^p)^2} \sum_{j=1}^{n} |\frac{\partial f}{\partial z_j}(z)|^2 dm_{2n}(z) < +\infty,$$

then f has a meromorphic extension f^* to Ω. If $p = 0$, then f^* is holomorphic.

Proof. The case $p = 0$ is a part of the cited result of [51], Theorem 3.5, pp. 300-301. Thus we may suppose that $p \neq 0$. Consider first the case in which $n = 1$.

An easy computation shows that

$$\int_{\mathbb{C}} \frac{|w|^{p-2}}{(1+|w|^p)^2} dm_2(w) = \frac{2\pi}{|p|}.$$

Take $z_0 \in E$ arbitrarily. By assumption, there is an $r_0 = r_0(p, f, z_0) > 0$ such that

$$\int_{B^2(z_0,r_0)\setminus F} \frac{|f(z)|^{p-2}}{(1+|f(z)|^p)^2} |f'(z)|^2 dm_2(z) < \frac{\pi}{|p|}.$$

Since clearly

$$\int_{\mathbb{C}} \frac{|w|^{p-2}}{(1+|w|^p)^2} dm_2(w) \leq \int_{B^2(z_0,r_0)\setminus F} \frac{|f(z)|^{p-2}}{(1+|f(z)|^p)^2} |f'(z)|^2 dm_2(z) < \frac{\pi}{|p|},$$

$f|B^2(z_0, r_0) \setminus E$ omits a set of positive measure. Thus by a result of Kametani [76], Theorem 2, p. 10, and Remark, p. 11, f has a meromorphic extension to $B^2(z_0, r_0)$.

The case in which $n \geq 2$ is proved using Fubini's theorem, Lemma 9.1 (i) and the following lemma. □

Lemma 11.17. *Suppose that Ω is a domain in $\mathbb{C}^n$, $n \geq 2$, $F \subset \Omega$ is closed in Ω and f is holomorphic in $\Omega \setminus F$. If for each j, $1 \leq j \leq n$, and for $\mathcal{H}^{2n-2}$-almost all*

$Z_j \in \mathbb{C}^{n-1}$, *the section* $F(Z_j)$ *is totally disconnected and the holomorphic function* $f_{Z_j} : (\Omega \setminus F)(Z_j) \to \mathbb{C}$

$$f_{Z_j}(z_j) = f(z_j, Z_j),$$

has a meromorphic extension to $\Omega(Z_j)$, *then* f *has a meromorphic extension to* Ω.

Remark 11.18. For related, partly previous and partly more general results, see [13], Theorem, p. 284, [51], Theorem 3.5, pp. 300-301, and [95], Theorem 3.1, pp. 925-926. As a matter of fact, we prove the following more general result.

Corollary 11.19. ([77], Théorème 20, p. 182, [50], Theorem 3.4, p. 477) *Suppose that* Ω *is a domain in* $\mathbb{C}^n$, $n \geq 1$, *and that* $F \subset \Omega$ *is closed in* Ω *and polar in* $\mathbb{R}^{2n}$. *Suppose that* f *is holomorphic in* $\Omega \setminus E$. *If the (pluri)subharmonic function* $\log^+ |f|$ *has a harmonic majorant in* $\Omega \setminus F$, *then* f *has a meromorphic extension to* Ω.

Proof. Because of the inequality $\log(1+x^2) \leq 2\log^+ |x| + \log 2$, the (pluri)subharmonic function $\log(1+|f|^2)$ has a harmonic majorant in $\Omega \setminus F$. Proceeding then as in [93], proof of Corollary 1, p. 549, one sees that the measure $\mu = \Delta \log(1+|f|^2)$ has *locally finite mass near* E, i.e. for each domain D relatively compact in Ω, one has $\mu(D \setminus F) < +\infty$. See also [13], p. 283. On the other hand, computing the Laplacian, one easily gets

$$\mu(D \setminus F) = 4 \int_{\Omega \setminus F} \frac{1}{(1+|f(z)|^2)^2} \sum_{j=1}^{n} |\frac{\partial f}{\partial z_j}(z)|^2 \, dm_{2n}(z)$$

for each domain D relatively compact in Ω. Compare [51], p. 297, [93], p. 549, and [157], p. 402. Since $\mu(D \setminus F) < +\infty$, the result follows then from Theorem 11.11. □

Remark 11.20. Observe that the above Corollary gives a positive answer to a question posed by Cima and Graham [16], Remarks 7.4, p. 255. See also [50], p. 470.

Next an extension result for meromorphic functions:

Theorem 11.21. ([91] Theorem 3.1, p. 147) *Suppose that* Ω *is a domain in* $\mathbb{C}^n$, $n \geq 1$, *and* $F \subset \Omega$ *is closed in* Ω *and such that* $\mathcal{H}^{2n-1}(F) < \sigma_\infty$. *Let* $F_1 \subset F$ *such that* $\mathcal{H}^{2n-2}(F_1) = 0$. *Suppose that* $f : \Omega \setminus F_1 \to \mathbb{C}^*$ *is a spherically continuous function such that* $f|\Omega \setminus F$ *is holomorphic. Then* $f|\Omega \setminus F$ *has a unique meromorphic extension to* Ω.

Corollary 11.22. ([91] Corollary 3.2, p. 148) *Suppose that* Ω *is a domain in* $\mathbb{C}^n$, $n \geq 1$, *and* $F \subset \Omega$ *is closed in* Ω *and such that* $\mathcal{H}^{2n-2}(F) = 0$. *Suppose* f *is meromorphic in* $\Omega \setminus E$. *Then* f *has a unique meromorphic extension to* Ω.

11.5. **Removable singularities of holomorphic functions with locally finite Riesz mass.**

Proposition 11.23. *Let* f *be holomorphic in* $\Omega \setminus F$ *and let* $p > 0$ *and* $\mu = \Delta |f|^p$. *Then the conditions*

(i) $\mu(D \setminus F) < +\infty$ *for each open set* $D \subset\subset \Omega$,

(ii) $\int_{D\setminus F} |f(z)|^{p-2} \sum_{j=1}^{n} \left|\frac{\partial f}{\partial z_j}(z)\right|^2 dm_{2n}(z) < +\infty$ *for each open set* $D \subset\subset \Omega$,

are equivalent, where in (ii) in the case when $p < 2$ *the integrand is defined as* 0 *in every component of* $D \setminus F$ *in which* f *vanishes identically. Moreover, (i) implies the condition*

(iii) f *belongs to the Hardy class* $\mathcal{H}^p(D\setminus F)$ *for each open set* $D \subset\subset \Omega$.

If F *is polar (in* $\mathbb{R}^{2n}$*), then (i), (ii) and (iii) are equivalent.*

Proof. Because of [13] p. 283, and [50] Lemma 3.3, p. 476, it is sufficient to show that

$$p^2 \int_{D\setminus F} |f(z)|^{p-2} \sum_{j=1}^{n} \left|\frac{\partial f}{\partial z_j}(z)\right|^2 dm_{2n}(z) = \mu(D\setminus F) \tag{11.6}$$

for each open set $D \subset\subset \Omega$. By an easy computation of the Laplacian, one sees that (11.6) holds when $p \geq 2$. In the general case one must show that $\mu(Z(f) \cap D) = 0$, where $Z(f) = \{z \in \Omega \setminus F : f(z) = 0\}$. We may assume that $Z(f)$ does not have interior points. Let D_1 be an open set such that $\overline{D_1} \subset D \setminus E$. Using the Riesz decomposition theorem (see for example [46], Theorem 5.25, p. 251) in the set D_1, we see that the Green potential of the measure $\mu|D_1$ is bounded in D_1. Since the set $Z(f)$ is polar, we have $\mu(Z(f) \cap D) = 0$. □

Remark 11.24. In general, (iii) does not imply (i). This can be seen, for example, as follows. In $\mathbb{C}$ there are sets of linear measure zero which are not removable singularities for functions in the Hardy classes (see for example [47]. However such sets are removable singularities for functions satisfying (i) or (ii), by Corollary 11.10 above.

Remark 11.25. If $n = 1$ and $p = 2$, the condition (ii) just says that f has finite Dirichlet integral locally near F.

12. QUASINEARLY SUBHARMONIC FUNCTIONS IN LOCALLY UNIFORMLY HOMOGENEOUS SPACES

Abstract. We define locally uniformly homogeneous spaces, and consider quasinearly subharmonic functions there. Especially, we consider weighted boundary behavior and boundary integral inequalities of quasinearly subharmonic functions.

Keywords. Locally uniformly homogeneous space, quasinearly subharmonic, weighted boundary behavior, boundary integral inequalities

12.1. **Locally uniformly homogeneous spaces.** The definition of locally uniformly homogeneous spaces was given in [86]. However, for the convenience of the reader we recall it here, too. A set X is *a locally uniformly homogeneous space* if the following conditions are satisfied:

(i) X is a topological space.
(ii) There is a Borel measure μ defined on X.
(iii) There is a *quasimetric* (*quasidistance*) on X, that is, there is a constant $K \geq 1$ and a mapping $d_K : X \times X \to [0, +\infty)$ such that
  1° $d_K(x,y) = d_K(y,x)$ for all $x, y \in X$,
  2° $d_K(x,y) = 0$ if and only if $x = y$,
  3° $d_K(x,y) \leq K[d_K(x,z) + d_K(z,y)]$ for all $x, y, z \in X$,
  4° the ((K-)quasi)balls $B_K(x,r)$,

$$B_K(x,r) := \{y \in X : d_K(x,y) < r\},$$

  centered at x and of radii $r > 0$, form a basis of open neighborhoods at the point $x \in X$,
  5° $0 < \mu(B_K(x,r)) < +\infty$ for all $x \in X$ and $r > 0$,
  6° there exist absolute constants $A = A(K) \geq 1$ and $\rho_0 = \rho_0(K) > 0$ such that

$$\mu(B_K(x,r)) \leq A\mu(B_K(x,\frac{r}{2}))$$

  for all $x \in X$ and all r, $0 < r \leq \rho_0$.

Remark 12.1. Locally uniformly homogeneous spaces are slightly more general than spaces of homogeneous type, defined and considered by Coifman and Weiss [17], pp. 66–68, and [18], pp. 587–590. As a matter of fact, the only difference with their definition is that, instead of the above condition 6°, Coifman and Weiss use the stronger condition:

$6'^o$ There exists an absolute constant $A = A(K) \geq 1$ such that

$$\mu(B_K(x,r)) \leq A\mu(B_K(x,\frac{r}{2}))$$

for all $x \in X$ and all $r > 0$.

For a list of examples of spaces of homogeneous type, see [18], pp. 588–590.

Juhani Riihentaus

Remark 12.2. In order to be able to consider Hausdorff measures on locally uniformly homogeneous spaces, we make the following additional assumption (cf. [71], p. 54): Let X be a locally uniformly homogeneous space. Suppose that X satisfies the following additional condition:

$$\text{(12.1)} \quad \begin{aligned} &\textit{For every } \delta > 0 \textit{ there are } E_j \in \mathcal{F}, j = 1,2,\ldots, \textit{ such that} \\ &X = \cup_{j=1}^{+\infty} E_j \textit{ and } d_K(E_j) \leq \delta, \textit{where } \mathcal{F} = \{\, B_K(x,r) \,:\, x \in X, r > 0 \,\}. \end{aligned}$$

Then for each $d > 0$ one can define in X a d-dimensional Hausdorff (outer) measure $\mathcal{H}_K^d$, which is a (K-quasi)metric (outer) measure in the following sense: If $A, B \subset X$ such that $d_K(A,B) > 0$, then $\mathcal{H}_K^d(A \cup B) = \mathcal{H}_K^d(A) + \mathcal{H}_K^d(B)$. As a matter of fact, in the standard definition (see e.g. [69], pp. 125–126), just work with the quasimetric d_K instead of the metric d (or ρ). One sees also, that all Borel sets of X are $\mathcal{H}_K^d$-measurable. Above we have used the following notation: If $A, B \subset X$, then

$$d_K(A) := \sup\{\, d_K(x,y) \,:\, x,y \in A \,\} \quad \text{and} \quad d_K(A,B) := \inf\{\, d_K(x,y) \,:\, x \in A,\, y \in B \,\}.$$

12.2. **Quasinearly subharmonic functions.** Though the definition of quasinearly subharmonic functions in locally uniformly homogeneous spaces was given in [86], we recall it also here for the convenience of the reader. Let X be a locally uniformly homogeneous space. Let $u : X \to [0, +\infty)$ be Borel measurable. Let $C \geq 1$. Then u is *C-quasinearly subharmonic in X* if there is a constant $\varepsilon_0 = \varepsilon_0(u)$ (depending on the considered function u), $0 < \varepsilon_0 < 1$, such that for each open set $\Omega \subset X$, $\Omega \neq X$, for each $x \in \Omega$ and each r, $0 < r \leq \min\{\, \rho_0,\, \varepsilon_0 \delta_K^{\Omega}(x) \,\}$, one has $u \in \mathcal{L}^1(B_K(x,r))$ and

$$u(x) \leq \frac{C}{\mu(B_K(x,r))} \int\limits_{B_K(x,r)} u(y)\, d\mu(y).$$

The function u is *quasinearly subharmonic in X* if u is C-quasinearly subharmonic for some $C \geq 1$. Above (and below) we have used the following notation: $\delta_K^{\Omega}(x)$, or shortly $\delta_K(x)$, is the (K-)quasidistance from $x \in \Omega$ to $\partial\Omega$, and thus defined by

$$\delta_K(x) := \delta_K^{\Omega}(x) := \inf\{\, d_K(x,y) \,:\, y \in \Omega^c \,\}$$

where Ω^c is the complement of Ω, taken in X.

12.3. **Examples.** Quasinearly subharmonic functions, especially nearly subharmonic, quasisubharmonic and subharmonic functions in an open subset D of an Euclidean space $\mathbb{R}^n$, $n \geq 2$, give examples of quasinearly subharmonic functions in a locally uniformly homogeneous space. As an additional example, we recall that B^{2n}, the unit ball of $\mathbb{C}^n$, $n \geq 1$, is locally uniformly homogeneous, and nonnegative $\mathcal{M}$-subharmonic functions on B^{2n} (see e.g. [138], p. 31, and [139], p. 3774) are 1-quasinearly subharmonic. For further examples, see [86].

For the definition, examples and properties of quasinearly subharmonic functions (sometimes, however, perhaps with a different terminology) in domains of an Euclidean space $\mathbb{R}^n$, $n \geq 2$, see e.g. [80], pp. 18–19, [81], pp. 15–16, [97], p. 233, [100], p. 171, [101], pp. 196–197, [104], p. 28, [105], p. 158, [60], pp. 243–244, [109], p. 52, [85], pp. 90–91, [110], pp. 2–3, [111], p. 2614, [112], pp. 129–130, [29], pp. 2–6, [30], and the references therein. In this connection, see also [151], pp. 259, 263.

12.4. **Weighted boundary behavior.** The following theorem is a special case of the original result of Gehring [39], Theorem 1, p. 77, and of Hallenbeck [43], Theorems 1 and 2, pp. 117–118, and of the later and more general results of Stoll [140], Theorems 1 and 2, pp. 301–302, 307:

Theorem 12.3. *If u is a function harmonic in the unit disk $\mathbb{D}$ of the complex plane $\mathbb{C}$ such that*

$$I(u) := \int_{\mathbb{D}} |\, u(z)\, |^p \,(1-|\, z\,|)^{\beta}\, dm_2(z) < +\infty, \tag{12.2}$$

where $p > 0$, $\beta > -1$, then

$$\lim_{r\to 1-} |\, u(re^{i\theta})\, |^p \,(1-r)^{\beta+1} = 0 \tag{12.3}$$

for almost all $\theta \in [0, 2\pi)$. Above m_2 is the Lebesgue measure in $\mathbb{R}^2$.

Observe that Gehring, Hallenbeck and Stoll considered in fact subharmonic functions and that the limit in (12.3) was uniform in Stolz approach regions, in Stoll's result in even more general regions. For more general results, see [97], Theorem, p. 233, [75], Theorem 2, p. 73, [100], Theorem 2, pp. 175–176, [101], Theorem 3.4.1, pp. 198–199, [104], Theorem, p. 31, and [85], Theorem 4, p. 102.

Gehring's proof was based on Hardy-Littlewood inequality, whereas the other authors based their proofs, more or less, on certain generalized mean value inequalities for subharmonic functions. For such inequalities and related properties, see [36], Lemma 2, p. 172, [64], Theorem 1, p. 529, [149], pp. 188-190, [78], pp. 53, 64–65, [94], Lemma, p. 69, [43], Lemma 1, p. 113, [75], p. 68, [99], Theorem, p. 188, and the references therein.

With the aid of the following Theorem 12.4, see [83], Theorem 1, pp. 433–434, Pavlović showed that the convergence in (12.3) is dominated. At the same time he pointed out that whole Theorem 12.3 follows from his result:

Theorem 12.4. *If u is a function harmonic in $\mathbb{D}$ satisfying (12.2), where $p > 0$, $\beta > -1$, then*

$$J(u) := \int_0^{2\pi} \sup_{0<r<1} |\, u(re^{i\theta})\, |^p \,(1-r)^{\beta+1}\, d\theta < +\infty.$$

Moreover, there is a constant $C = C_{p,\beta}$ such that $J(u) \leq C I(u)$.

In [112], Theorems 1 and 2, pp. 131–132, we extended Theorem 12.4 to the case, where, instead of absolute values of harmonic functions in the unit disk $\mathbb{D}$ of the complex plane $\mathbb{C}$, one considers more generally nonnegative quasinearly subharmonic functions in rather general domains of $\mathbb{R}^n$, $n \geq 2$. Now our aim is to extend this cited Theorem 1 even further: We will give a related result for quasinearly subharmonic functions in locally uniformly homogeneous spaces, satisfying the above additional assumption (12.1), see Theorem 12.9 below. As an application, we get in Corollary 12.13 below a weighted boundary behavior result in our rather general setup of locally uniformly homogeneous spaces.

12.5. **Admissible functions.** A function $\varphi : [0,+\infty) \to [0,+\infty)$ is *admissible*, if it is strictly increasing, surjective, and there are constants $C_0 = C_0(\varphi) \geq 1$ and $r_0 = r_0(\varphi) > 0$ such that

$$\varphi(2t) \leq C_0\,\varphi(t) \quad \text{and} \quad \varphi^{-1}(2s) \leq C_0\,\varphi^{-1}(s) \tag{12.4}$$

for all s, t, $0 \leq s, t \leq r_0$.

Examples of admissible functions are: Functions $\varphi_1(t) = t^p$, $p > 0$, nonnegative, increasing surjective functions $\varphi_2(t)$ satisfying the Δ_2-condition and for which the functions $t \mapsto \frac{\varphi_2(t)}{t}$ are increasing, and functions $\varphi_3(t) = ct^{\alpha}[\log(\delta + t^{\gamma})]^{\beta}$, where $c > 0$, $\alpha > 0$, $\delta \geq 1$, and $\beta, \gamma \in \mathbb{R}$ are such that $\alpha + \beta\gamma > 0$.

12.6. **Approach sets.** Let $\varphi : [0,+\infty) \to [0,+\infty)$ be an admissible function and let $\alpha > 0$. Let X be a locally uniformly homogeneous space. Let Ω be a domain in a component X_1 of X, $\Omega \neq X_1$. For $\zeta \in \partial\Omega$ write

$$\Gamma_{\varphi}(\zeta,\alpha) := \{\, x \in \Omega \, : \, \varphi(d_K(x,\zeta)) < \alpha\,\delta_K(x) \,\},$$

and call it a (φ,α)-*approach set (region)*, shortly an *approach set (region), in* Ω *at* ζ. Observe that though $\partial\Omega$ is surely nonempty, the approach set $\Gamma_{\varphi}(\zeta,\alpha)$ may, in certain cases, be empty. Anyway, in the case of the unit disk $\mathbb{D}$ of the complex plane $\mathbb{C}$, the choice $\varphi(t) = t$ gives the familiar Stolz approach regions. Choosing $\varphi(t) = t^{\tau}$, $\tau \geq 1$, say, one gets more general approach regions, see [140], p. 301.

For $x \in \Omega$ and $\alpha > 0$, we also write

$$\tilde{\Gamma}_{\varphi}(x,\alpha) := \{\, \xi \in \partial\Omega \, : \, x \in \Gamma_{\varphi}(\xi,\alpha) \,\}.$$

Moreover, for $\rho > 0$ we write

$$\Gamma_{\varphi,\rho}(\zeta,\alpha) := \{\, x \in \Gamma_{\varphi}(\zeta,\alpha) \, : \, \delta_K(x) < \rho \,\}.$$

One says that $\zeta \in \partial\Omega$ is (φ,α)-*accessible*, shortly *accessible*, if $\Gamma_{\varphi}(\zeta,\alpha) \cap B_K(\zeta,\rho) \neq \emptyset$ for all $\rho > 0$.

12.7. **Ahlfors-regular sets.** Let $d > 0$. A set $E \subset X$ is *Ahlfors-regular from above, with dimension d and with constant* $C_4 > 0$, shortly *Ahlfors-regular from above*, if it is closed and

$$\mathcal{H}^d_K(E \cap B_K(x,r)) \leq C_4 r^d$$

for all $x \in E$ and $r > 0$. The smallest constant C_4 is called the *regularity constant* for E. A set $E \subset X$ is *Ahlfors-regular, with dimension d and with constant* $C_4 > 0$, shortly *Ahlfors-regular*, if it is closed and

$$C_4^{-1} r^d \leq \mathcal{H}^d_K(E \cap B_K(x,r)) \leq C_4 r^d$$

for all $x \in E$ and $r > 0$.

Simple examples of Ahlfors-regular sets in $\mathbb{R}^n$, $n \geq 2$, are d-planes and d-dimensional Lipschitz graphs. Also certain Cantor sets and self-similar sets are Ahlfors-regular. For more details, see [21], pp. 9–10.

12.8. **Boundary integral inequalities.** We begin with four lemmas. Recall that X is always a locally uniformly homogeneous space. X_1 will be an arbitrary component of X, and Ω a domain in X_1, $\Omega \neq X_1$. Moreover $\varphi : [0,+\infty) \to [0,+\infty)$ is an admissible function, with constants r_0 and C_0.

Let $x \in \Omega$. It is easy to see that

$$\frac{2}{3K}\delta_K(x) \leq \delta_K(y) \leq (K+\frac{1}{3})\delta_K(x)$$

for all $y \in B_K(x)$. Let $\alpha > 0$. Write

$$\hat{\rho}_0 := \min\{\frac{r_0}{2^{K+\frac{1}{3}}}, \frac{r_0}{2^{\alpha+1}}, \frac{r_0}{2^{\frac{3\alpha K}{2}C_0^{K+\frac{1}{3}}+1}(K+\frac{1}{3})}, \frac{1}{2^{\alpha+1}}\varphi\left(\frac{r_0}{2^{K+\frac{1}{3}}}\right), \rho_0\}.$$

Lemma 12.5. *Let $\zeta \in \partial\Omega$ and $x_0 \in \Gamma_{\varphi,\rho}(\zeta,\alpha)$. Let $C_1 \geq 1$ be arbitrary. Then for $C_2' = \frac{C_0^\alpha}{3} + KC_0^{K+\frac{1}{3}}$ and for all $x \in B_K(x_0)$,*

$$B_K(x, C_1\varphi^{-1}(\delta_K(x))) \subset B_K(x_0, C_1C_2'\varphi^{-1}(\delta_K(x_0))),$$

provided $0 < \rho \leq \hat{\rho}_0$.

Proof. Take $z \in B_K(x, C_1\varphi^{-1}(\delta_K(x)))$ arbitrarily. Then

$$\begin{aligned}
d_K(x_0,z) &\leq K[d_K(x_0,x) + d_K(x,z)] < \\
&< K[\frac{\delta_K(x_0)}{3K} + C_1\varphi^{-1}(\delta_K(x))] < \\
&< K[\frac{\delta_K(x_0)}{3K} + C_1\varphi^{-1}((K+\frac{1}{3})\delta_K(x_0))] < \\
&< K[\frac{\delta_K(x_0)}{3K} + C_1\varphi^{-1}(2^{K+\frac{1}{3}}\delta_K(x_0))] < \\
&< K[\frac{\delta_K(x_0)}{3K} + C_1C_0^{K+\frac{1}{3}}\varphi^{-1}(\delta_K(x_0))] < \\
&< K\left(\frac{C_0^\alpha}{3K} + C_1C_0^{K+\frac{1}{3}}\right)\varphi^{-1}(\delta_K(x_0)) \leq \\
&\leq C_1C_2'\varphi^{-1}(\delta_K(x_0)),
\end{aligned}$$

where $C_2' = \frac{C_0^\alpha}{3} + KC_0^{K+\frac{1}{3}}$. Hence $z \in B_K(x_0, C_1C_2'\varphi^{-1}(\delta_K(x_0)))$. Above we have used the facts that $2^{K+\frac{1}{3}}\hat{\rho}_0 \leq r_0$ and $2^{\alpha+1}\hat{\rho}_0 \leq r_0$, which follow from the definition of $\hat{\rho}_0$. □

Lemma 12.6. *Let $\zeta \in \partial\Omega$ and $x_0 \in \Gamma_{\varphi,\rho}(\zeta,\alpha)$. Then $B_K(x_0) \subset \Gamma_{\varphi,\rho'}(\zeta,\alpha')$, where $\rho' = (K+\frac{1}{3})\rho$ and $\alpha' = \frac{3\alpha K}{2}C_0^{K+\frac{1}{3}}$, provided $0 < \rho \leq \hat{\rho}_0$.*

Proof. Take $x \in B_K(x_0)$ arbitrarily. Then $d_K(x_0,x) < \frac{\delta_K(x_0)}{3K}$. Since $\varphi(d_K(x_0,\zeta)) < \alpha\delta_K(x_0)$, we have

$$\begin{aligned}\varphi(d_K(x,\zeta)) &< \varphi(K[\frac{\delta_K(x_0)}{3K} + d_K(x_0,\zeta)]) \leq \varphi(K[\frac{d_K(x_0,\zeta)}{3K} + d_K(x_0,\zeta)]) < \\ &< \varphi((K+\frac{1}{3})d_K(x_0,\zeta)) \leq \varphi(2^{K+\frac{1}{3}}d_K(x_0,\zeta)) < \\ &< C_0^{K+\frac{1}{3}}\varphi(d_K(x_0,\zeta)) < C_0^{K+\frac{1}{3}}\alpha\delta_K(x_0) \leq \frac{C_0^{K+\frac{1}{3}}3\alpha K}{2}\delta_K(x),\end{aligned}$$

provided that $2^{K+\frac{1}{3}}d_K(x_0,\zeta) \leq r_0$. But this surely holds, since $x_0 \in \Gamma_{\varphi,\rho}(\zeta,\alpha)$ and $\rho \leq \hat{\rho}_0 \leq \frac{1}{2^{\alpha+1}}\varphi\left(\frac{r_0}{2^{K+\frac{1}{3}}}\right)$. Hence $x \in \Gamma_{\varphi,\rho'}(\zeta,\alpha')$. □

Lemma 12.7. *Let* $C_1' = C_0^{\frac{3\alpha K}{2}C_0^{K+\frac{1}{3}}+1}$ *and* $\alpha' = \frac{3\alpha K}{2}C_0^{K+\frac{1}{3}}$. *Then for all* $x \in \Omega_{\rho'}$, *where* $\rho' = (K+\frac{1}{3})\rho$, *one has* $\tilde{\Gamma}_\varphi(x,\alpha') \subset B_K(x, C_1'\varphi^{-1}(\delta_K(x)))$, *provided* $0 < \rho \leq \hat{\rho}_0$.

Proof. Suppose that $\tilde{\Gamma}_\varphi(x,\alpha') \neq \emptyset$ and take $\xi \in \tilde{\Gamma}_\varphi(x,\alpha')$ arbitrarily. But then $x \in \Gamma_\varphi(\xi,\alpha')$, that is $\varphi(d_K(x,\xi)) < \alpha'\delta_K(x)$. Hence

$$d_K(x,\xi) < \varphi^{-1}(\alpha'\delta_K(x)) < \varphi^{-1}(2^{\alpha'+1}\delta_K(x)) \leq C_0^{\alpha'+1}\varphi^{-1}(\delta_K(x)),$$

provided that $2^{\alpha'+1}(K+\frac{1}{3})\hat{\rho}_0 \leq r_0$. But this holds, since, by assumption,

$$\hat{\rho}_0 \leq \frac{r_0}{2^{\frac{3\alpha K}{2}C_0^{K+\frac{1}{3}}+1}(K+\frac{1}{3})}.$$

Thus $\xi \in B_K(x, C_1'\varphi^{-1}(\delta_K(x)))$. □

Lemma 12.8. *Let* $\zeta \in \partial\Omega$ *and* $x \in \Gamma_{\varphi,\rho}(\zeta,\alpha)$. *Let* C_1' *and* C_2' *be as above. Then for* $C_3' = K\left(1 + \frac{C_0^{\alpha+1}}{C_1'C_2'}\right)$ *one has*

$$B_K(x, C_1'C_2'\varphi^{-1}(\delta_K(x))) \subset B_K(\zeta, C_1'C_2'C_3'\varphi^{-1}(\delta_K(x))),$$

provided $0 < \rho \leq \hat{\rho}_0$.

Proof. Take $z \in B_K(x, C_1'C_2'\varphi^{-1}(\delta_K(x)))$ arbitrarily. Then clearly

$$d_K(z,\zeta) \leq K[d_K(x,z) + d_K(x,\zeta)] < K[C_1'C_2'\varphi^{-1}(\delta_K(x)) + d_K(x,\zeta)].$$

Now $x \in \Gamma_{\varphi,\rho}(\zeta,\alpha)$, thus $\varphi(d_K(x,\zeta)) < \alpha\delta_K(x)$, and also

$$d_K(x,\zeta) < \varphi^{-1}(\alpha\delta_K(x)) \leq \varphi^{-1}(2^{\alpha+1}\delta_K(x)) \leq C_0^{\alpha+1}\varphi^{-1}(\delta_K(x)),$$

provided that $2^{\alpha+1}\hat{\rho}_0 \leq r_0$, which again holds, since, by assumption, $\hat{\rho}_0 \leq \frac{r_0}{2^{\alpha+1}}$. Hence,

$$d_K(z,\zeta) < K[C_1'C_2'\varphi^{-1}(\delta_K(x)) + C_0^{\alpha+1}\varphi^{-1}(\delta_K(x))] <$$
$$< K(C_1'C_2' + C_0^{\alpha+1})\varphi^{-1}(\delta_K(x)) = C_1'C_2'K\left(1 + \frac{C_0^{\alpha+1}}{C_1'C_2'}\right)\varphi^{-1}(\delta_K(x)),$$

and so $z \in B_K(\zeta, C_1'C_2'C_3'\varphi^{-1}(\delta_K(x)))$. □

Then our result, an extension to Theorem 1, pp. 131-132, of [112]:

Theorem 12.9. *Let X be a locally uniformly homogeneous space satisfying the condition (12.1). Suppose that $d_K : X \times X \to [0,+\infty)$ is separately continuous and a Borel function. Let $\varphi : [0,+\infty) \to [0,+\infty)$ be an admissible function, with constants r_0 and C_0. Let $\alpha > 0$, $\gamma \in \mathbb{R}$, $d > 0$ and $C_4 > 0$ be arbitrary. Let $u : X \to [0,+\infty)$ be a K_1-quasinearly subharmonic function. Then there is a constant $C = C(\alpha,\gamma,\varepsilon_0,d,A,C_0,C_4,K,K_1)$ such that for each component X_1 of X and for each domain $\Omega \subset X_1$, $\Omega \neq X_1$, whose boundary $\partial\Omega$ is Ahlfors-regular from above, with dimension d and with constant C_4,*

(12.5)
$$\int_{\partial\Omega} \sup_{x\in\Gamma_{\varphi,\rho}(\zeta,\alpha)} \{\delta_K(x)^\gamma \mu(B_K(x))[\varphi^{-1}(\delta_K(x))]^{-d}u(x)\}\, d\mathcal{H}_K^d(\zeta) \leq C \int_{\Omega_{\rho'}} \delta_K(x)^\gamma u(x)\, d\mu(x),$$

for all ρ, $0 < \rho \leq \hat{\rho}_0$. Here $\rho' = (K + \frac{1}{3})\rho$ and $\hat{\rho}_0$ is as above.

Remark 12.10. Above and below we use the following (maybe unstandard, but in the considered situation of nonnegative functions nevertheless natural) convention: Let $A \subset X$, $B \subset A$ and $g : A \to [0,+\infty]$. If $B = \emptyset$, then we *define* $\sup_{x\in B}\{g(x)\} = 0$.

Remark 12.11. Though the constant C above in (12.5) does depend on K_1 and on ε_0, it is, nevertheless, otherwise independent of the K_1-quasinearly subharmonic function u.

Proof. Suppose $0 < \rho \leq \hat{\rho}_0$. Write

$$E := \{\zeta \in \partial\Omega : \Gamma_{\varphi,\rho}(\zeta,\alpha) \neq \emptyset\}.$$

Using the fact that $d_K(\cdot,\cdot)$ is separately upper semicontinuous, one sees easily that E is open in $\partial\Omega$.

Take $\zeta \in E$ and $x_0 \in \Gamma_{\varphi,\rho}(\zeta,\alpha)$ arbitrarily. Since u is quasinearly subharmonic and

$$\frac{\varepsilon_0\delta_K(x_0)}{3K} < \varepsilon_0\delta_K(x_0) < \delta_K(x_0) \leq \rho \leq \hat{\rho}_0 \leq \rho_0$$

one obtains

$$u(x_0) \leq \frac{K_1}{\mu(B_K(x_0, \frac{\varepsilon_0 \delta_K(x_0)}{3K}))} \int\limits_{B_K(x_0, \frac{\varepsilon_0 \delta_K(x_0)}{3K})} u(x)\, d\mu(x) \leq$$

$$\leq \frac{K_1}{\mu(B_K(x_0, \frac{\varepsilon_0 \delta_K(x_0)}{3K}))} \int\limits_{B_K(x_0, \frac{\delta_K(x_0)}{3K})} u(x)\, d\mu(x).$$

Choose $n_0 \in \mathbb{N}$ such that

$$2^{n_0 - 1} < \frac{1}{\varepsilon_0} \leq 2^{n_0}.$$

Then

$$\mu(B_K(x_0, \frac{\delta_K(x_0)}{3K})) = \mu(B_K(x_0, \frac{1}{\varepsilon_0} \cdot \frac{\varepsilon_0 \delta_K(x_0)}{3K})) \leq \mu(B_K(x_0, 2^{n_0} \cdot \frac{\varepsilon_0 \delta_K(x_0)}{3K})) \leq$$

$$\leq A^{n_0} \mu(B_K(x_0, \frac{\varepsilon_0 \delta_K(x_0)}{3K})) \leq A^{1 - \log_2 \varepsilon_0} \mu(B_K(x_0, \frac{\varepsilon_0 \delta_K(x_0)}{3K})).$$

Hence
(12.6)

$$u(x_0) \leq \frac{K_1 A^{1 - log_2 \varepsilon_0}}{\mu(B_K(x_0, \frac{\delta_K(x_0)}{3K}))} \int\limits_{B_K(x_0, \frac{\delta_K(x_0)}{3K})} u(x)\, d\mu(x) = \frac{K_1 A^{1 - log_2 \varepsilon_0}}{\mu(B_K(x_0))} \int\limits_{B_K(x_0)} u(x)\, d\mu(x).$$

With the aid of the fact that $\delta_K(x_0)^\gamma \leq (\frac{3K}{2})^{|\gamma|} \delta_K(x)^\gamma$ for all $x \in B_K(x_0)$ and that $[\varphi^{-1}(\delta_K(x_0))]^d \geq \frac{1}{C_0^{d(K+\frac{1}{3})}} [\varphi^{-1}(\delta_K(x))]^d$ for all $x \in B_K(x_0)$, we get, with the aid of Lemma 12.5 above, from (12.6) above, for all $C_1 \geq 1$,

$$\frac{\delta_K(x_0)^\gamma \mu(B_K(x_0)) u(x_0)}{[\varphi^{-1}(\delta_K(x_0))]^d + \mathcal{H}_K^d(B_K(x_0, C_1 C_2' \varphi^{-1}(\delta_K(x_0))) \cap \partial\Omega)} \leq$$

$$\leq K_1 A^{1 - log_2 \varepsilon_0} \int\limits_{B_K(x_0)} \frac{\delta_K(x_0)^\gamma u(x)}{[\varphi^{-1}(\delta_K(x_0))]^d + \mathcal{H}_K^d(B_K(x_0, C_1 C_2' \varphi^{-1}(\delta_K(x_0))) \cap \partial\Omega)} d\mu(x) \leq$$

$$\leq K_1 A^{1 - log_2 \varepsilon_0} \int\limits_{B_K(x_0)} \frac{(\frac{3K}{2})^{|\gamma|} \delta_K(x)^\gamma u(x)}{\frac{1}{C_0^{d(K+\frac{1}{3})}} [\varphi^{-1}(\delta_K(x))]^d + \mathcal{H}_K^d(B_K(x, C_1 \varphi^{-1}(\delta_K(x))) \cap \partial\Omega)} d\mu(x) \leq$$

$$\leq \left(\frac{3K}{2}\right)^{|\gamma|} K_1 A^{1 - log_2 \varepsilon_0} C_0^{d(K+\frac{1}{3})} \int\limits_{B_K(x_0)} \frac{\delta_K(x)^\gamma u(x)}{[\varphi^{-1}(\delta_K(x))]^d + \mathcal{H}_K^d(B_K(x, C_1 \varphi^{-1}(\delta_K(x))) \cap \partial\Omega)} d\mu(x).$$

By Lemma 12.6, $B_K(x_0) \subset \Gamma_{\varphi,\rho'}(\zeta,\alpha')$, where $\rho' = (K+\frac{1}{3})\rho$ and $\alpha' = \frac{3\alpha K}{2}C_0^{K+\frac{1}{3}}$. Thus

$$\frac{\delta_K(x_0)^\gamma \mu(B_K(x_0))u(x_0)}{[\varphi^{-1}(\delta_K(x_0))]^d + \mathcal{H}_K^d(B_K(x_0,C_1C_2'\varphi^{-1}(\delta_K(x_0)))\cap\partial\Omega)} \leq$$
$$\leq \left(\frac{3K}{2}\right)^{|\gamma|} K_1 A^{1-\log_2\varepsilon_0} C_0^{d(K+\frac{1}{3})} \int\limits_{\Gamma_{\varphi,\rho'}(\zeta,\alpha')} \frac{\delta_K(x)^\gamma u(x)}{[\varphi^{-1}(\delta_K(x))]^d + \mathcal{H}_K^d(B_K(x,C_1\varphi^{-1}(\delta_K(x)))\cap\partial\Omega)} d\mu(x).$$

Taking then the supremum on the left hand side over $x_0 \in \Gamma_{\varphi,\rho}(\zeta,\alpha)$, we get

$$\sup_{x_0\in\Gamma_{\varphi,\rho}(\zeta,\alpha)} \frac{\delta_K(x_0)^\gamma \mu(B_K(x_0))u(x_0)}{[\varphi^{-1}(\delta_K(x_0))]^d + \mathcal{H}_K^d(B_K(x_0,C_1C_2'\varphi^{-1}(\delta_K(x_0)))\cap\partial\Omega)} \leq$$
$$\leq \left(\frac{3K}{2}\right)^{|\gamma|} K_1 A^{1-\log_2\varepsilon_0} C_0^{d(K+\frac{1}{3})} \int\limits_{\Gamma_{\varphi,\rho'}(\zeta,\alpha')} \frac{\delta_K(x)^\gamma u(x)}{[\varphi^{-1}(\delta_K(x))]^d + \mathcal{H}_K^d(B_K(x,C_1\varphi^{-1}(\delta_K(x)))\cap\partial\Omega)} d\mu(x).$$

Next integrate on both sides with respect to ζ over E and use Fubini's theorem:

$$\int\limits_E \sup_{x_0\in\Gamma_{\varphi,\rho}(\zeta,\alpha)} \frac{\delta_K(x_0)^\gamma \mu(B_K(x_0))u(x_0)}{[\varphi^{-1}(\delta_K(x_0))]^d + \mathcal{H}_K^d(B_K(x_0,C_1C_2'\varphi^{-1}(\delta_K(x_0)))\cap\partial\Omega)} d\mathcal{H}_K^d(\zeta) \leq$$
$$\leq \left(\frac{3K}{2}\right)^{|\gamma|} K_1 A^{1-\log_2\varepsilon_0} C_0^{d(K+\frac{1}{3})} \int\limits_E \{ \int\limits_{\Gamma_{\varphi,\rho'}(\zeta,\alpha')} \frac{\delta_K(x)^\gamma u(x)}{[\varphi^{-1}(\delta_K(x))]^d + \mathcal{H}_K^d(B_K(x,C_1\varphi^{-1}(\delta_K(x)))\cap\partial\Omega)} d\mu(x)\} d\mathcal{H}_K^d(\zeta) \leq$$
$$\leq \left(\frac{3K}{2}\right)^{|\gamma|} K_1 A^{1-\log_2\varepsilon_0} C_0^{d(K+\frac{1}{3})} \int\limits_E \{ \int\limits_{\Omega_{\rho'}} \chi_{\Gamma_\varphi(\zeta,\alpha')}(x) \frac{\delta_K(x)^\gamma u(x)}{[\varphi^{-1}(\delta_K(x))]^d + \mathcal{H}_K^d(B_K(x,C_1\varphi^{-1}(\delta_K(x)))\cap\partial\Omega)} d\mu(x)\} d\mathcal{H}_K^d(\zeta) \leq$$
$$\leq \left(\frac{3K}{2}\right)^{|\gamma|} K_1 A^{1-\log_2\varepsilon_0} C_0^{d(K+\frac{1}{3})} \int\limits_E \{ \int\limits_{\Omega_{\rho'}} \chi_{\tilde{\Gamma}_\varphi(x,\alpha')}(\zeta) \frac{\delta_K(x)^\gamma u(x)}{[\varphi^{-1}(\delta_K(x))]^d + \mathcal{H}_K^d(B_K(x,C_1\varphi^{-1}(\delta_K(x)))\cap\partial\Omega)} d\mu(x)\} d\mathcal{H}_K^d(\zeta) \leq$$
$$\leq \left(\frac{3K}{2}\right)^{|\gamma|} K_1 A^{1-\log_2\varepsilon_0} C_0^{d(K+\frac{1}{3})} \int\limits_{\Omega_{\rho'}} \{ \int\limits_E \chi_{\tilde{\Gamma}_\varphi(x,\alpha')}(\zeta) d\mathcal{H}_K^d(\zeta)\} \frac{\delta_K(x)^\gamma u(x)}{[\varphi^{-1}(\delta_K(x))]^d + \mathcal{H}_K^d(B_K(x,C_1\varphi^{-1}(\delta_K(x)))\cap\partial\Omega)} d\mu(x) \leq$$
$$\leq \left(\frac{3K}{2}\right)^{|\gamma|} K_1 A^{1-\log_2\varepsilon_0} C_0^{d(K+\frac{1}{3})} \int\limits_{\Omega_{\rho'}} \mathcal{H}_K^d(\tilde{\Gamma}_\varphi(x,\alpha')) \frac{\delta_K(x)^\gamma u(x)}{[\varphi^{-1}(\delta_K(x))]^d + \mathcal{H}_K^d(B_K(x,C_1\varphi^{-1}(\delta_K(x)))\cap\partial\Omega)} d\mu(x).$$

Choosing $C_1 = C_1'$ and using then Lemma 12.8 we get

$$\int_E \sup_{x_0\in\Gamma_{\varphi,\rho}(\zeta,\alpha)} \frac{\delta_K(x_0)^\gamma \mu(B_K(x_0))u(x_0)}{[\varphi^{-1}(\delta_K(x_0))]^d + \mathcal{H}_K^d(B_K(x_0, C_1'C_2'\varphi^{-1}(\delta_K(x_0)))\cap\partial\Omega)} d\mathcal{H}_K^d(\zeta) \leq$$

$$\leq \left(\frac{3K}{2}\right)^{|\gamma|} K_1 A^{1-log_2\varepsilon_0} C_0^{d(K+\frac{1}{3})} \int_{\Omega_{\rho'}} \frac{\mathcal{H}_K^d(B_K(x, C_1'\varphi^{-1}(\delta_K(x)))\cap\partial\Omega)\delta_K(x)^\gamma u(x)}{[\varphi^{-1}(\delta_K(x))]^d + \mathcal{H}_K^d(B_K(x, C_1'\varphi^{-1}(\delta_K(x)))\cap\partial\Omega)} d\mu(x)$$

$$\leq \left(\frac{3K}{2}\right)^{|\gamma|} K_1 A^{1-log_2\varepsilon_0} C_0^{d(K+\frac{1}{3})} \int_{\Omega_{\rho'}} \delta_K(x)^\gamma u(x)\, d\mu(x).$$

On the other hand, by Lemma 12.8, we get

$$\sup_{x\in\Gamma_{\varphi,\rho}(\zeta,\alpha)} \frac{\delta_K(x)^\gamma \mu(B_K(x))u(x)}{[\varphi^{-1}(\delta_K(x))]^d + \mathcal{H}_K^d(B_K(x, C_1'C_2'\varphi^{-1}(\delta_K(x)))\cap\partial\Omega)} \geq$$

$$\geq \sup_{x\in\Gamma_{\varphi,\rho}(\zeta,\alpha)} \frac{\delta_K(x)^\gamma \mu(B_K(x))u(x)}{[\varphi^{-1}(\delta_K(x))]^d + \mathcal{H}_K^d(B_K(\zeta, C_1'C_2'C_3'\varphi^{-1}(\delta_K(x)))\cap\partial\Omega)}.$$

Since $\partial\Omega$ is Ahlfors-regular from above, one has

$$\mathcal{H}_K^d(B_K(\zeta, C_1'C_2'C_3'\varphi^{-1}(\delta_K(x)))\cap\partial\Omega) \leq C_4[C_1'C_2'C_3'\varphi^{-1}(\delta_K(x))]^d < +\infty.$$

Therefore

$$\sup_{x\in\Gamma_{\varphi,\rho}(\zeta,\alpha)} \frac{\delta_K(x)^\gamma \mu(B_K(x))u(x)}{[\varphi^{-1}(\delta_K(x))]^d + \mathcal{H}_K^d(B_K(\zeta, C_1'C_2'C_3'\varphi^{-1}(\delta_K(x)))\cap\partial\Omega)} \geq$$

$$\geq \sup_{x\in\Gamma_{\varphi,\rho}(\zeta,\alpha)} \frac{\delta_K(x)^\gamma \mu(B_K(x))u(x)}{[\varphi^{-1}(\delta_K(x))]^d + C_4[C_1'C_2'C_3'\varphi^{-1}(\delta_K(x))]^d} =$$

$$= \frac{1}{1 + C_4(C_1'C_2'C_3')^d} \sup_{x\in\Gamma_{\varphi,\rho}(\zeta,\alpha)} \{\delta_K(x)^\gamma \mu(B_K(x))[\varphi^{-1}(\delta_K(x))]^{-d} u(x)\}.$$

Thus we have:

(12.7)

$$\int_E \sup_{x\in\Gamma_{\varphi,\rho}(\zeta,\alpha)} \{\delta_K(x)^\gamma \mu(B_K(x))[\varphi^{-1}(\delta_K(x))]^{-d} u(x)\} d\mathcal{H}_K^d(\zeta) \leq C \int_{\Omega_{\rho'}} \delta_K(x)^\gamma u(x)\, d\mu(x),$$

where

$$C = \left(\frac{3K}{2}\right)^{|\gamma|} K_1 A^{1-log_2\varepsilon_0} C_0^{d(K+\frac{1}{3})} [1 + C_4\left(C_1'C_2'C_3'\right)^d]$$

and

$$C_1' = C_0^{\frac{3\alpha K}{2} C_0^{K+\frac{1}{3}}+1}, \quad C_2' = \frac{C_0^\alpha}{3} + K C_0^{K+\frac{1}{3}}, \quad C_3' = K\left(1 + \frac{C_0^{\alpha+1}}{C_1'C_2'}\right).$$

To conclude the proof, observe the following. First, since $\Gamma_{\varphi,\rho}(\zeta,\alpha)=\emptyset$ for all $\zeta\in\partial\Omega\setminus E$, we can, just using our convention in Remark 12.10, replace (12.7) by the desired inequality:

$$\int_{\partial\Omega}\sup_{x\in\Gamma_{\varphi,\rho}(\zeta,\alpha)}\{\delta_K(x)^\gamma\mu(B_K(x))[\varphi^{-1}(\delta_K(x))]^{-d}u(x)\}d\mathcal{H}_K^d(\zeta)\leq C\int_{\Omega_{\rho'}}\delta_K(x)^\gamma u(x)\,d\mu(x).$$

Second, the functions

$$\partial\Omega\ni\zeta\mapsto\sup_{x\in\Gamma_{\varphi,\rho}(\zeta,\alpha)}\frac{\delta_K(x)^\gamma\mu(B_K(x))u(x)}{[\varphi^{-1}(\delta_K(x))]^d+\mathcal{H}_K^d(B_K(x,C_1C_2'\varphi^{-1}(\delta_K(x)))\cap\partial\Omega)}\in[0,+\infty]$$

and

$$\partial\Omega\ni\zeta\mapsto\sup_{x\in\Gamma_{\varphi,\rho}(\zeta,\alpha)}\{\delta_K(x)^\gamma\mu(B_K(x))[\varphi^{-1}(\delta_K(x))]^{-d}u(x)\}\in[0,+\infty]$$

are lower semicontinuous. Thus the above integrations on "the left hand sides" are justified.

Third, the functions

$$\Omega_{\rho'}\ni x\mapsto\frac{\delta_K(x)^\gamma u(x)}{[\varphi^{-1}(\delta_K(x))]^d+\mathcal{H}_K^d(B_K(x,C_1\varphi^{-1}(\delta_K(x)))\cap\partial\Omega)}\in[0,+\infty)$$

and

$$\Omega_{\rho'}\times\partial\Omega\ni(x,\zeta)\mapsto\chi_{\Gamma_\varphi(\zeta,\alpha')}(x)\frac{\delta_K(x)^\gamma u(x)}{[\varphi^{-1}(\delta_K(x))]^d+\mathcal{H}_K^d(B_K(x,C_1\varphi^{-1}(\delta_K(x)))\cap\partial\Omega)}=$$
$$=\chi_{\tilde{\Gamma}_\varphi(x,\alpha')}(\zeta)\frac{\delta_K(x)^\gamma u(x)}{[\varphi^{-1}(\delta_K(x))]^d+\mathcal{H}_K^d(B_K(x,C_1\varphi^{-1}(\delta_K(x)))\cap\partial\Omega)}\in[0,+\infty)$$

are Borel measurable. Hence the integrations and the use of Fubini's theorem on "the right hand sides" are justified, too. Observe that here we use our additional assumption that the K-quasimetric d_K is Borel measurable. □

Remark 12.12. At present we do not know whether our assumption that the K-quasimetric $d_K:X\times X\to[0,+\infty)$ is separately continuous and Borel measurable, is really necessary or not. Observe anyway that a quasimetric d_K is separately upper semicontinuous, see 12.1 (iii) 4° above. But a separately upper semicontinuous function need not, however, be measurable, see [136] and e.g. [41], Example 1, p. 11. On the other hand, if $K=1$, that is, if X is a metric space, and the function d_1 is separately continuous, then d_1 is Borel measurable by a result of Kuratowski, see [128], p. 742, and the references therein, say. Observe also that if the considered locally uniformly homogeneous space X is moreover locally compact, then by a result of Johnson, see [55], Theorem 2.2, p. 422, and again [128], p. 742, d_K is indeed Borel measurable.

Corollary 12.13. *Let X be a locally uniformly homogeneous space satisfying the condition (12.1). Suppose that $d_K:X\times X\to[0,+\infty)$ is separately continuous and a Borel function. Let $\varphi:[0,+\infty)\to[0,+\infty)$ be an admissible function, with constants r_0 and C_0. Let $\alpha>0$, $\gamma\in\mathbb{R}$, $d>0$ and $C_4>0$ be arbitrary. Let $u:X\to[0,+\infty)$ be a*

K_1-quasinearly subharmonic function. Suppose that X_1 is an arbitrary component of X and that $\Omega \subset X_1$, $\Omega \neq X_1$, is a domain, whose boundary $\partial\Omega$ is Ahlfors-regular from above, with dimension d and with constant C_4, and that

$$\int_{\Omega} \delta_K(x)^{\gamma} u(x)\, d\mu(x) < +\infty. \tag{12.8}$$

Then for $\mathcal{H}_K^d$-almost every (φ,α)-accessible point $\zeta \in \partial\Omega$,

$$\lim_{\rho\to 0}\left(\sup_{x\in\Gamma_{\varphi,\rho}(\zeta,\alpha)} \{\delta_K(x)^{\gamma}\mu(B_K(x))[\varphi^{-1}(\delta_K(x))]^{-d}u(x)\}\right) = 0.$$

Remark 12.14. If, instead of a locally uniformly homogeneous space X, one works in an Euclidean space $\mathbb{R}^n$, $n \geq 2$, then slightly better results hold: Namely, one can omit the assumed Ahlfors-regularity condition of $\partial\Omega$, see the already cited results [97], Theorem, p. 233, [100], Theorem 2, pp. 175–176, [102], Theorem 3.4.1, pp. 198–199, [104], Theorem, p. 31, and [85], Theorem 4, p. 102. Observe, however, that the possibility for this omission in the Euclidean setup is based essentially on a well-known density estimate result for Hausdorff measures (which in turn is based, among others, on Vitali's covering theorem and thus is of "very Euclidean space-type"), see e.g. [71], Theorem 6.2, p. 89.

Proof. By Theorem 12.9,

$$\int_{\partial\Omega} \sup_{x\in\Gamma_{\varphi,\rho}(\zeta,\alpha)} \{\delta_K(x)^{\gamma}\mu(B_K(x))[\varphi^{-1}(\delta_K(x))]^{-d}u(x)\}\, d\mathcal{H}_K^d(\zeta) \leq C \int_{\Omega_{\rho'}} \delta_K(x)^{\gamma}u(x)\, d\mu(x),$$

where $C = C(\alpha,\gamma,\varepsilon_0,d,A,C_0,C_4,K,K_1)$. Using then just Fatou's lemma and (12.8), one sees that

$$\int_{\partial\Omega} \liminf_{\rho\to 0}\left(\sup_{x\in\Gamma_{\varphi,\rho}(\zeta,\alpha)} \{\delta_K(x)^{\gamma}\mu(B_K(x))[\varphi^{-1}(\delta_K(x))]^{-d}u(x)\}\right) d\mathcal{H}_K^d(\zeta) \leq$$

$$\leq C \liminf_{\rho\to 0} \int_{\Omega_{\rho'}} \delta_K(x)^{\gamma}u(x)\, d\mu(x) = 0.$$

Thus the claim follows. □

NOTATION

Our notation is rather standard, see e.g. [49, 85, 86]. However, for the convenience of the reader we recall the following. The common convention $0 \cdot (\pm\infty) = 0$ is used. Below D and Ω are usually domains of either of $\mathbb{R}^n$ or of $\mathbb{C}^n$, when $n \geq 1$. X is always a locally uniformly homogeneous space, and when $\Omega \subset X$, then Ω is a domain in X such that $\Omega = X$, whose boundary $\partial\Omega$ is Ahlfors-regular from above, with dimension $d > 0$ and with constant $C_4 > 0$ (for the definition of this see 12.7 above). For $\rho > 0$ write $\Omega_\rho = \{x \in \Omega : \delta_K(x) < \rho\}$. $B_K(x,r)$ is the ((K-)quasi)ball in X, with center x and radius r, and $B_K(x) = B_K(x, \frac{1}{3K}\delta_K(x))$. The d-dimensional Hausdorff (outer) measure in X, $d > 0$, constructed with the aid of the K-quasimetric d_K, is denoted by $\mathcal{H}^d_K$, see Remark 12.2 above. C_0 and r_0 are fixed constants which are involved with the used (and thus fixed) admissible function φ (see (12.4) above in 12.5). Similarly, if $\alpha > 0$ is given, $C_1' = C_1'(\alpha, C_0, K)$, $C_2' = C_2'(\alpha, C_0, K)$ and $C_3' = C_3'(\alpha, C_0)$ are fixed constants, coming directly from Lemmas 12.5, 12.7 and 12.8 above. (Compare these with the related constants C_1, C_2 and C_3 in [97], p. 234, [104], pp. 32–33, and [112], p. 129.)

REFERENCES

[1] P. Ahern, J. Bruna, "Maximal and area integral characterizations of Hardy-Sobolev spaces in the unit ball of $\mathbb{C}^n$", *Revista Mat. Iberoamericana*, vol. 4, 123–153, 1988.

[2] P. Ahern, W. Rudin, "Zero sets of functions in harmonic Hardy spaces", *Math. Scand.*, vol. 73, 209–214, 1993.

[3] D.H. Armitage, S.J. Gardiner, "Conditions for separately subharmonic functions to be subharmonic", *Potential Anal.*, vol. 2, 255–261, 1993.

[4] D.H. Armitage, S.J. Gardiner, *Classical Potential Theory*, Springer-Verlag, London, 2001.

[5] M.G. Arsove, "On subharmonicity of doubly subharmonic functions", *Proc. Amer. Math. Soc.*, vol. 17, 622–626, 1966.

[6] V. Avanissian, "Fonctions plurisousharmoniques et fonctions doublement sousharmoniques", *Ann. Sci. École Norm. Sup.*, vol. 78, 101–161, 1961.

[7] V. Avanissian, "Sur l'harmonicité des fonctions séparément harmoniques", in: *Séminaire de Probabilités (Univ. Strasbourg, Février 1967)*, vol. 1, 3–17, 1966/1967, Springer, Berlin, 1967.

[8] A.S. Besicovitch, "On sufficient conditions for a function to be analytic, and on behavior of analytic functions in the neighborhood of non-isolated singular point", *Proc. London Math. Soc. (2)*, vol. 32, 1–9, 1931.

[9] P. Blanchet, "On removable singularities of subharmonic and plurisubharmonic functions", *Complex Variables*, vol. 26, 311–322, 1995.

[10] P. Blanchet, P.M. Gauthier, "Fusion of two solutions of a partial differential equation", *Meth, Appl. Anal.*, vol. 1, no. 3, 371–384, 1994.

[11] M. Brelot, "Minorantes sous-harmoniques, extrémales et capacités", *J. Math. Pures Appl.*, vol. 24,1–32, 1945.

[12] Yu. Burago, V. Zalgaller, *Geometric Inequalities*, Springer-Verlag, New York, 1988.

[13] U. Cegrell, "Removable singularity sets for analytic functions having modulus with bounded Laplace mass", *Proc. Amer. Math. Soc.*, vol. 88, 283–286, 1983.

[14] U. Cegrell, A. Sadullaev, "Separately subharmonic functions", *Uzbek. Math. J.*, vol. 1, 78–83, 1993.

[15] E.M. Chirka, *Complex Analytic Sets*, Kluwer Academic Publisher, Dordrecht, 1989.

[16] J. Cima, I. Graham, "On the extension of holomorphic functions with growth conditions across analytic subvarieties", *Michigan Math, J.*, vol. 28, 241–256, 1981.

[17] Ronald R. Coifman, Guido Weiss, *Analyse Harmonique Non-commutative Sur Certains Espaces Homogènes*, Lecture Notes in Mathematics, vol. 242, Springer-Verlag, Berlin · Heidelberg · New York, 1971.

[18] Ronald R. Coifman, Guido Weiss, "Extensions of Hardy Spaces and Their Use in Analysis", *Bull. Amer. Math. Soc.*, vol. 83, no. 1, 569–645, 1977.

[19] B.J. Cole, T.J. Ransford, "Subharmonicity without upper semicontinuity", *J. Functional Anal.*, vol. 147, 420–442, 1997.

[20] R.O. Davies, H. Fast, "Lebesgue density influences Hausdorff measure; Large sets surface-like from many directions", *Mathematika*, vol. 25, 116–119, 1978.

[21] G. David, S. Semmes, *Analysis of and on Uniformly Rectifiable Sets*, Mathematical Surveys and Monographs, vol. 38, American Mathematical Society, Providence, Rhode Island, 1991.

[22] E. Di Benedetto, N.S. Trudinger, "Harnack inequalities for quasi-minima of variational integrals", *Ann. Inst. H. Poincaré, Analyse Nonlineaire*, vol. 1, 295–308, 1984.

[23] J. Dieudonné, *Foundations of Modern Analysis*, Academic Press, New York, 1960.

[24] J. Dieudonné, *Algèbre linéaire et géometrie élémentaire*, troisiéme ed., Hermann, Paris, 1968.

[25] O. Djordjević, M. Pavlović, "Equivalent norms on Dirichlet spaces of polyharmonic functions on the ball in $\mathbb{R}^N$", *Bol. Soc. Mat. Mexicana (3)*, vol. 13, 307–319, 2007.

[26] O. Djordjević, M. Pavlović, "$\mathcal{L}^p$-integrability of the maximal function of a polyharmonic function", *J. Math. Anal. Appl.*, vol. 336, 411–417, 2007.

Juhani Riihentaus

[27] Y. Domar, "On the existence of a largest subharmonic minorant of a given function", *Arkiv mat.*, vol. 3, no. 39, 429–440, 1957.

[28] Y. Domar, "Uniform boundedness in families related to subharmonic functions", *J. London Math. Soc. (2)*, vol. 38, 485–491, 1988.

[29] O. Dovgoshey, J. Riihentaus, "Bi-Lipschitz mappings and quasinearly subharmonic functions", *Int. J. Math. Math. Sci.*, vol. 2010, 1–8, 2010.

[30] O. Dovgoshey, J. Riihentaus, "A remark concerning generalized mean value inequalities for subharmonic functions", International Conference Analytic Methods of Mechanics and Complex Analysis, Dedicated to N.A. Kilchevskii and V.A. Zmorovich on the Occasion of their Birthday Centenary, Kiev, Ukraine, June 29 - July 5, 2009, in: *Transactions of the Institute of Mathematics of the National Academy of Ukraine*, vol. 7, no. 2, 26–33, 2010.

[31] O. Dovgoshey, J. Riihentaus, "Mean type inequalities for quasinearly subharmonic functions", *Glasgow Math. J.*, vol. 55, 349–368, 2013.

[32] O. Dovgoshey, J. Riihentaus, "On quasinearly subharmonic functions", Dedicated to Professor Vladimir Gutlyanskii on the Occasion of his 75-th Anniversary – *Lobachevskii Journal of Mathematics*, vol. 38, no. 2, 245-254, 2017. (doi: 10.1134/S1995080217020068)

[33] Lawrence C. Evans, Ronald F. Gariepy, *Measure Theory and Fine Properties of Functions*, Studies in Advanced Mathematics, CRC Press, Boca Raton, Ann Arbor, London, 1992.

[34] K.J. Falconer, *Fractal Geometry*, John Wiley & Sons, 1993.

[35] H. Federer, *Geometric Measure Theory*, Springer-Verlag, Berlin, 1969.

[36] C. Fefferman, E.M. Stein, "H^p spaces of several variables", *Acta Math.*, vol. 129, 137–192, 1972.

[37] S.J. Gardiner, "Removable singularities for subharmonic functions", *Pac. J. Math.*, vol. 147, 71–80, 1991.

[38] J.B. Garnett, *Bounded Analytic Functions*, Springer-Verlag, New York, 2007 (Revised First Edition).

[39] F.W. Gehring, "On the radial order of subharmonic functions", *J. Math. Soc. Japan*, vol. 9, 77–79, 1957.

[40] F.W. Gehring, O. Martio, "Quasiextremal distance domains and extension of quasiconformal mappings", *J. d'Analyse Mathématique*, vol. 45, 181–206, 1985.

[41] Zbigniew Grande, "Quasicontinuity and measurability of functions of two variables", *Real Analysis Exchange*, vol. 28, 7–14, 2002.

[42] J.W. Green, "Approximately subharmonic functions", *Proc. Amer. Math. Soc.*, vol. 3, 829–833, 1952.

[43] D.J. Hallenbeck, "Radial growth of subharmonic functions", in: *Pitman Research Notes*, vol. 262, 113–121, 1992.

[44] R. Harvey, J. Polking, "Removable singularities of solutions of linear partial differential equations", *Acta Math.*, vol. 125, 39–56, 1970.

[45] R. Harvey, J. Polking, "Extending analytic objects", *Comm. Pure and Appl. Mathematics*, vol. 28, 701–727, 1975.

[46] W.K. Hayman, P.B. Kennedy, *Subharmonic Functions, I*, Academic Press, 1976.

[47] D.A. Hejhal, "Classification theory for Hardy classes of analytic functions", *Ann. Acad. Sci. Fenn. Ser. A I*, vol. 566, 1–29, 1974.

[48] L.L. Helms, *Introduction to Potential Theory*, Wiley-Interscience, New York, 1969.

[49] M. Hervé, *Analytic and Plurisubharmonic Functions in Finite and Infinite Dimensional Spaces*, Lecture Notes in Mathematics 198, Springer-Verlag, Berlin, 1971.

[50] J. Hyvönen, J. Riihentaus, "On the extension in the Hardy classes and in the Nevanlinna class", *Bull. Soc. Math. France*, vol. 112, 469–480, 1984.

[51] J. Hyvönen, J. Riihentaus, "Removable singularities for holomorphic functions with locally finite Riesz mass", *J. London Math. Soc. (2)*, vol. 35, 296–302, 1987.

[52] S.A. Imomkulov, "Separately subharmonic functions" (in Russian), *Dokl. USSR*, vol. 2, 8–10, 1990.

[53] M. Jarnicki, M. Pflug, *Extension of Holomorphic Functions*, Walter de Gruyter, Berlin, 2000.

[54] M. Jarnicki, M. Pflug, *Separately Analytic Functions*, European Mathematical Society, Zürich, 2011.

[55] B.E. Johnson, "Separate continuity and measurability", *Proc. Amer. Math. Soc.*, vol. 20, 420–422, 1969.

[56] M. Kalantar, "Families of subharmonic functions and separately subharmonic functions", *Comp. Var. Ell. Eq.*, vol. 65(5), 886-895, 2020. (https://doi.org/ 10.1080/17476933.2019.1652280)

[57] R. Kaufman, J.-M. Wu, "Removable singularities for analytic or subharmonic functions", *Ark. mat.*, vol. 18, 109–116, 1980.

[58] B.N. Khabibullin, "A uniqueness theorem for subharmonic functions of finite order", *Mat. Sb.*, vol. 182(6), 811–827, 1991; English transl., in: *Math. USSR Sbornik*, vol. 73(1), 195–210, 1992.

[59] B.N. Khabibullin, "Completeness of systems of entire functions in spaces of holomorphic functions", *Mat. Zametki*, vol. 66(4), 603–616, 1999; English transl., in: *Math. Notes*, vol. 66(4), 495–506, 1999.

[60] V. Kojić, "Quasi-nearly subharmonic functions and conformal mappings", *Filomat.*, vol. 21, no. 2, 243–249, 2007.

[61] S. Kołodziej, J. Thorbiörnson, J., "Separately harmonic and subharmonic functions", *Potential Anal.*, vol. 5, 463–466, 1996.

[62] P. Koskela, "Removable singularities for analytic functions", *Michigan Math. J.*, vol. 40, 459–466, 1993.

[63] P. Koskela, V. Manojlović, "Quasi-nearly subharmonic functions and quasiconformal mappings", *Potential Anal.*, vol. 37, 187–196, 2012.

[64] Ü. Kuran, "Subharmonic behavior of $| h |^p$, ($p > 0$, h harmonic)", *J. London Math. Soc. (2)*, vol. 8, 529–538, 1974.

[65] P. Lelong, "Les fonctions plurisousharmoniques", *Ann. Sci. École Norm. Sup.*, vol. 62, 301–338, 1945.

[66] P. Lelong, "Fonctions plurisousharmoniques et fonctions analytiques de variables réelles", *Ann. Inst. Fourier, Grenoble*, vol. 11, 515–562, 1961.

[67] P. Lelong, *Plurisubharmonic Functions and Positive Differential Forms*, Gordon and Breach, London, 1969.

[68] E.H. Lieb, M. Loss, *Analysis*, Graduate Studies in Mathematics, vol. 14, American Mathematical Society, Providence, Rhode Island, 2001.

[69] Jaroslav Lukeš, Jan Malý, *Measure and Integral.* Matfyzpress, Publishing House of the Faculty of Mathematics and Physics, Charles University, Prague, 1995.

[70] P. Mattila, "An inequality for capacities", *Math. Scand.*, vol. 53, 256–264, 1983.

[71] P. Mattila, *Geometry of Sets and Measures in Euclidean Spaces*, Cambridge studies in advanced mathematics, vol. 44, Cambridge University Press, Cambridge, 1995.

[72] Olivera R. Mihić, "Some properties of quasinearly subharmonic functions and maximal theorem for Bergman type spaces", *ISRN Mathematical Analysis*, vol. 2013, Article ID 515398, 3 pages, 2013. (http://dx.doi.org/10.1155/2013/515398)

[73] Y. Mizuta, "Boundary limits of harmonic functions in Sobolev-Orlicz classes", in: *Potential Theory* (ed. M. Kishi), Walter de Gruyter & Co, Berlin · New York, 1991, pp. 235–249.

[74] Y. Mizuta, *Potential Theory in Euclidean Spaces*, Gaguto International Series, Mathematical Sciences and Applications, vol. 6, Gakkōtosho Co., Tokyo, 1996.

[75] Y. Mizuta, "Boundary limits of functions in weighted Lebesgue or Sobolev classes", *Revue Roum. Math. Pures Appl.*, vol. 46, 67–75, 2001.

[76] K. Noshiro, *Cluster Sets*, Springer-Verlag, Berlin · Göttingen · Heidelberg, 1960.

[77] M. Parreau, "Sur les moyennes des fonctions harmoniques et analytiques et la classification des surfaces de Riemann", *Ann. Inst. Fourier (Grenoble)*, vol. 3, 103–197, 1951.

[78] M. Pavlović, "Mean values of harmonic congugates in the unit disc", *Complex Variables*, vol. 10, 53–65, 1988.

[79] M. Pavlović, "Inequalities for the gradient of eigenfunctions of the invariant Laplacian on the unit ball", *Indag. Math. (N.S.)*, vol 2, 89–98, 1991.

[80] M. Pavlović, "On subharmonic behavior and oscillation of functions on balls in $\mathbb{R}^n$", *Publ. Inst. Math. (Beograd)*, vol. 55 (69), 18–22, 1994.

[81] M. Pavlović, "Subharmonic behavior of smooth functions", *Mat. Vesnik*, vol. 48, 15–21, 1996.

[82] M. Pavlović, *Introduction to Function Spaces on the Disk*, Posebna Izdanja, vol. 20, Matematički Institut SANU, 2004.

[83] M. Pavlović, "An inequality related to the Gehring-Hallenbeck theorem on radial limits of functions in harmonic Bergman spaces", *Glasgow Math. J.*, vol. 50, no. 3, 433–435, 2008.

[84] M. Pavlović, *Function Classes on the Unit Disk*, Studies in Mathematics, vol. 52, Walter De Gruyter, Berlin, 2013, New York 2014.

[85] M. Pavlović, J. Riihentaus, "Classes of quasi-nearly subharmonic functions", *Potential Anal.*, vol. 29, 89–104, 2008.

[86] M. Pavlović, J. Riihentaus, "Quasi-nearly subharmonic functions in locally uniformly homogeneous spaces", *Positivity*, vol. 15, 1–10, 2011.

[87] J. Polking, "A survey of removable singularities", Seminar on nonlinear partial differential equations, Berkeley, California, 1983, in: *Math. Sci. Res. Inst. Publ.* (ed. S.S. Chern), vol. 2, 261–292, Springer-Verlag, Berlin, 1984.

[88] T. Radó, *Subharmonic Functions*, Springer-Verlag, Berlin, 1937.

[89] J. Riihentaus, "Removable singularities of analytic functions of several complex variables", *Math. Z.*, vol. 32, 45–54, 1978.

[90] J. Riihentaus, "Removable singularities of analytic and meromorphic functions of several complex variables", *Colloquim on Complex Analysis, Joensuu, Finland, August 24–27, 1978 (Complex Analysis, Joensuu 1978), in: Proceedings (eds. Ilpo Laine, Olli Lehto, Tuomas Sorvali), Lecture Notes in Mathematics*, vol. 747, 329–342, Springer-Verlag, Berlin, 1979.

[91] J. Riihentaus, "An extension theorem for meromorphic functions of several variables", *Ann. Acad. Sci. Fenn. Ser, A, I, Mathematica*, vol. 4, 145–149, 1978/1979.

[92] J. Riihentaus, "On the extension of separately hyperharmonic functions and H^p-functions", *Michigan Math. J.*, vol. 31, 99–112, 1984.

[93] J. Riihentaus, "Removable singularities in the Nevanlinna class and in the Hardy classes", *Proc. Amer. Math. Soc.*, vol. 102(3), 546–550, 1988.

[94] J. Riihentaus, "On a theorem of Avanissian–Arsove", *Expo. Math.*, vol. 7, 69–72, 1989.

[95] J. Riihentaus, "A nullset for normal functions in several variables", *Proc. Amer. Math. Soc.*, vol. 110(4), 923–933, 1990.

[96] J. Riihentaus, "Removable singularities for Bloch and normal functions", *Czech. Math. J.*, vol. 43(118), 723–741, 1993.

[97] J. Riihentaus, "Subharmonic functions: non-tangential and tangential boundary behavior", in: *Function Spaces, Differential Operators and Nonlinear Analysis (FSDONA'99), Proceedings of the Syöte Conference 1999*, V. Mustonen, J. Rákosnik (eds.), Math. Inst., Czech Acad. Science, Praha, 2000, pp. 229–238. (ISBN 80-85823-42-X)

[98] J. Riihentaus, "Removable sets for subharmonic functions", *Pac. J. Math.*, vol. 194(1), 199-208, 2000.

[99] J. Riihentaus, "A generalized mean value inequality for subharmonic functions", *Expo. Math.*, vol. 19, 187-190, 2001.

[100] J. Riihentaus, "A generalized mean value inequality for subharmonic functions and applications", *arXiv:math.CA/0302261v2 1 Nov 2006.*

[101] J. Riihentaus, "Subharmonic functions, mean value inequality, boundary behavior, nonintegrability and exceptional sets", International Workshop on Potential Theory and Free Boundary Flows, Kiev, Ukraine, August 19-27, 2003, in: *Transactions of the Institute of Mathematics of the National Academy of Sciences of Ukraine*, vol. 1, no. 3, 169–191, 2004.

[102] J. Riihentaus, "Weighted boundary behavior and nonintegrability of subharmonic functions", in: *International Conference on Education and Information Systems: Technologies and Applications (EISTA'04), Orlando, Florida, USA, July 21-25, 2004, Proceedings*, M. Chang, Y-T. Hsia, F. Malpica, M. Suarez, A. Tremante, F. Welsch (eds.), vol. II, 2004, pp. 196–202. (ISBN 980-6560-11-6)

[103] J. Riihentaus, "An integrability condition and weighted boundary behavior of subharmonic and $\mathcal{M}$-subharmonic functions: a survey", *Int. J. Diff. Eq. Appl.*, vol. 10, 1–14, 2005.

[104] J. Riihentaus, "On the weighted boundary behavior of $\mathcal{M}$-subharmonic functions", *International Workshop on Potential Theory and free Boundary Flows, September 25-30, 2005, Kiev, Ukraine*, in: *Transactions of the Institute of Mathematics of the National Academy of Ukraine*, vol. 3, no. 4, 92–108, 2006.

[105] J. Riihentaus, "A weighted boundary limit result for subharmonic functions", *Advances in Algebra and Analysis*, vol. 1, no. 1, 27-38, 2006.

[106] J. Riihentaus, "Separately quasi-nearly subharmonic functions", in: *Complex Analysis and Potential Theory, Proceedings of the Conference Satellite to ICM 2006*, Tahir Aliyev Azeroğlu, Promarz M. Tamrazov (eds.), Gebze Institute of Technology, Gebze, Turkey, September 8-14, 2006, World Scientific, Singapore, 2007, pp. 156–165.

[107] J. Riihentaus, "On the subharmonicity of separately subharmonic functions", in: *Proceedings of the 11th WSEAS International Conference on Applied Mathematics (MATH'07), Dallas, Texas, USA, March 22-24, 2007*, Kleanthis Psarris, Andrew D. Jones (eds.), WSEAS, 2007, pp. 230-236. (IBSN 978-960-8457-60-7)

[108] J. Riihentaus, "On separately harmonic and subharmonic functions", *Int. J. Pure Appl. Math.*, vol. 35, no. 4, 435-446, 2007.

[109] J. Riihentaus, "Subharmonic functions, generalizations and separately subharmonic functions", *The XIV-th Conference on Analytic Functions, July 22-28, 2007, Chełm, Poland*, in: *Scientific Bulletin of Chełm, Section of Mathematics and Computer Science*, vol. 2, 49–76, 2007. (ISBN 978-83-61149-24-8) (arXiv:math/0610259v5 [math.AP] 8 Oct 2008)

[110] J. Riihentaus, "Quasi-nearly subharmonicity and separately quasi-nearly subharmonic functions", *J. Inequal. Appl.*, vol. 2008, Article ID 149712, 15 pages, 2008. (doi: 10.1155/2008/149712)

[111] J. Riihentaus, "Separately subharmonic functions and quasi-nearly subharmonic functions", in: *The 12th Worlds Multi-Conference on Systemics, Cybernetics and Informatics (WMSCI 2008), Orlando, Florida, USA, June 29th-July 2nd, 2008, Proceedings*, N. Callaos, W. Lesso, C.D. Zinn, J. Baralt, J. Boukachour, C. White, T. Marwala, F. Nelwamondo (eds.), vol. V, 2008, pp. 53–56. (ISBN-10: 1-934272-35-3)

[112] J. Riihentaus, "On an inequality related to the radial growth of subharmonic functions", *CUBO, A Mathematical Journal*, vol. 11, N° 4, 127–136, 2009.

[113] J. Riihentaus, "Subharmonic functions, generalizations, weighted boundary behavior, and separately subharmonic functions: A survey", *Fifth World Congress of Nonlinear Analysts (WCNA 2008), Orlando, Florida, USA, July 2-9, 2008*, in: *Nonlinear Analysis, Series A: Theory, Methods & Applications*, vol. 71, no. 12, 15 December 2009, pp. e2613–e26267. (doi: 10.1016/j.na.2009.05.077)

[114] J. Riihentaus, "On an inequality related to the radial growth of quasinearly subharmonic functions in locally uniformly homogeneous spaces", *J. Math. Sciences: Advances and Appl.*, vol. 6, no. 1, 17–40, 2010.

[115] J. Riihentaus, "Domination conditions for families of quasinearly subharmonic functions", *Int. J. Math. Math. J.*, vol. 2011, Article ID 729849, 9 pages, 2011. (doi: 10.1155/2011/729849)

[116] J. Riihentaus, "An inequality type condition for quasinearly subharmonic functions and applications", *Positivity VII, Zaanen Centennial Conference, Leiden, 22-26 July, 2013, Conference Proceedings* (eds. Marcel de Jeu, Ben de Pagter, Onno van Gaans, Mark Veraar), in: *Ordered Structures and Applications: Positivity VII, Trends in Mathematics, Springer International Publishing*, 2016, pp. 395-414 (doi: 10.1007/978-3-319-27842-1_25). (arXiv:1509.05829v1 [math.AP] 19 Sep 2015)

[117] J. Riihentaus, "Remarks on separately subharmonic functions", *International Conference Complex Analysis, Potential Theory and Applications, Dedicated to the Memory of Professor P.M. Tamrazov, Kiev, Ukraine, August 19-23, 2013*, in: *Transactions of the Mathematics of the National Academy of Sciences of Ukraine*, Special Issue to Commemorate Professor P.M. Tamrazov, vol. 10, no. 1, 362–371, 2013.

[118] J. Riihentaus, "On separately subharmonic and harmonic functions", *Complex Variables and Elliptic Equations*, vol. 59, no. 2, 149–161, 2013. (doi: 10.1080/17476933.2013.816845)

[119] J. Riihentaus, "Exceptional sets for subharmonic functions", *J. Basic & Applied Sciences*, vol. 11, 567–571, 2015.

[120] J. Riihentaus, "A removability result for holomorphic functions of several complex variables", *J. Basic & Applied Sciences*, vol. 12, 50–52, 2016.

[121] J. Riihentaus, "Removability results for subharmonic functions, for harmonic functions and for holomorphic functions", *Matematychni Studii*, vol. 46, no. 2, 152–158, 2016.

[122] J. Riihentaus, "A removability result for separately subharmonic functions", *Visnyk of the Lviv Univ. Series Mech. Math.*, vol. 2018, Issue 85, 60–65, 2019.

[123] J. Riihentaus, "Removability results for subharmonic functions, for separately subharmonic functions, for harmonic functions, for separately harmonic functions and for holomorphic functions, a survey", *arXiv:1607.07029v9 [math.AP] 19 Jul 2019.*

[124] J. Riihentaus, "On the domination conditions for families of quasinearly subharmonic functions", in: *Theory and Applications of Mathematical Science*, vol. 2, 2020 (eds. Charles Roberto Telles et al.), Book Publisher International, 2020, Chapter 9, pp. 126-137. (doi: 10.9734/bpi/tams/v2, ISBN 978-93-89816-39-6)

[125] P.J. Rippon, "Some remarks on largest subharmonic minorants", *Math. Scand.*, vol. 49, 128–132, 1981.

[126] W. Rudin, "Integral representation of continuous functions", *Trans. Amer. Math. Soc.*, vol. 68, 278–286, 1950.

[127] W. Rudin, *Real and Complex Analysis*, Tata McGraw-Hill, New Delhi, 1979.

[128] Walter Rudin, "Lebesgue's First Theorem", *Mathematical Analysis and Applications, Part B, pp. 741–747*, in: *Advances in Mathematics Supplementary Studies*, vol. 78, Academic Press, New York · London, 1981.

[129] A. Sadullaev, "Rational approximation and pluripolar sets", *Mat. Sb.*, vol. 119, 96–118, 1982. (Russian)

[130] A. Sadullaev, "On separately subharmonic functions (Lelong's problem)", *Ann. Fac. Sci. Toulouse*, vol. 20, no. Spécial, 183-187, 2011.

[131] S. Saks, "On the operators of Blaschke and Privaloff for subharmonic functions", *Rec. Math. (Mat. Sbornik)*, vol. 9 (51), 451–456, 1941.

[132] V.L. Shapiro, "Generalized laplacians", *Amer. J. Math.*, vol. 78, 497–508, 1956.

[133] V.L. Shapiro, "Removable sets for pointwise subharmonic functions", *Trans. Amer. Math. Soc.*, vol. 159, 369–380, 1971.

[134] V.L. Shapiro, "Subharmonic functions and Hausdorff measure", *J. Diff. Eq.*, vol. 27, 28–45, 1978.

[135] B. Shiffman, "On the removal of singularities of analytic sets", *Michigan Math. J.*, vol. 15, 111–120, 1968.
[136] W. Sierpiński, "Sur un problème concernant les ensembles mesurables superficiellement", *Fund. Math.*, vol 1, 112–115, 1920.
[137] N. Sjöberg, "Sur les minorantes sousharmoniques d'une fonction donnée", *Neuvième Congrès des mathematiciens Scandinaves, Helsingfors*, pp. 309–319, 1938.
[138] M. Stoll, *Invariant Potential Theory in the Unit Ball of* $\mathbb{C}^n$, London Mathematical Society Lecture Notes Series, Cambridge, 1994.
[139] M. Stoll, "Boundary limits and non-integrability of $\mathcal{M}$-subharmonic functions in the unit ball of $\mathbb{C}^n$ $(n \geq 1)$", *Trans. Amer. Math. Soc.*, vol. 349, 3773–3785, 1997.
[140] M. Stoll, "Weighted tangential boundary limits of subharmonic functions on domains in $\mathbb{R}^n$ $(n \geq 2)$", *Math. Scand.*, vol. 83, 300–308, 1998.
[141] M. Stoll, "Harmonic majorants for eigenfunctions of the Laplacian with finite Dirichlet integrals", *J. Math. Anal. Appl.*, vol. 274, no. 2, 788–811, 2002. (doi: 10.1016/S0022-247X(02)00364-5)
[142] M. Stoll, "On generalizations of the Littlewood-Paley inequalities to domains in $\mathbb{R}^n$, $(n \geq 2)$", *International Workshop on Potential Theory in Matsue, Shimane University, Matsue, Japan, August 23-28, 2004.* A revised version of the paper "The Littlewood-Paley inequalities for Hardy-Orlicz spaces of harmonic functions on domains in $\mathbb{R}^n$" has been published in *Advanced Studies in Pure Mathematics*, vol. 4, pp. 363–376, 2006.
[143] M. Stoll, "On the Littlewood-Paley inequalities for subharmonic functions on domains in $\mathbb{R}^n$", *Recent Advances in Harmonic Analysis and Applications* (eds. D. Bilyk et al.), Springer Proceedings in Mathematics & Statistics, vol. 25, part 2, pp. 357–383, 2013. (doi: 10.1007/978-1-4614-4565-4_28)
[144] M. Stoll, "Littlewood-Paley theory for subharmonic functions on the unit ball in $\mathbb{R}^N$", *J. Math. Anal. Appl.*, vol. 420, no. 1, 483–514, 2014. (http://dx.doi.org/10.1016/j.maa.2014.05.017)
[145] N. Suzuki, "Nonintegrability of harmonic functions in a domain", *Japan. J. Math.*, vol. 16, 269–278, 1990.
[146] N. Suzuki, "Nonintegrability of superharmonic functions", *Proc. Amer. Math. Soc.*, vol. 113, no. 1, 113–115, 1991.
[147] E. Szpilrajn, "Remarques sur les fonctions sousharmoniques", *Ann. Math.*, vol. 34, 588–594, 1933.
[148] B.S. Thomson, *Real Functions*, Lecture Notes in Mathematics, vol. 1170, Springer, Berlin · Heidelberg, 1985.
[149] A. Torchinsky, *Real-Variable Methods in Harmonic Analysis*, Academic Press, London, 1986.
[150] D.C. Ullrich, "Removable sets for harmonic functions", *Michigan Math. J.*, vol. 38, 467–473, 1991.
[151] M. Vuorinen, "On the Harnack constant and the boundary behavior of Harnack functions", *Ann. Acad. Sci. Fenn., Ser. A I, Math.*, vol. 7, 259–277, 1982.
[152] M. Vuorinen, *Conformal Geometry and Quasiregular Mappings*, Lecture Notes in Mathematics, vol. 1319, Springer-Verlag, Berlin · Heidelberg · New York, 1988.
[153] J. Väisälä, *Lectures on n-Dimensional Quasiconformal Mappings*, Lecture Notes in Mathematics, vol. 229, Springer-Verlag, Berlin · Heidelberg · New York, 1971.
[154] R. Webster, *Convexity*, Oxford University Press, Oxford, 1994.
[155] J. Wiegerinck, "Separately subharmonic functions need not be subharmonic", *Proc. Amer. Math. Soc.*, vol. 104, 770–771, 1988.
[156] J. Wiegerinck, R. Zeinstra, "Separately subharmonic functions: when are they subharmonic", in: *Proceedings of Symposia in Pure Mathematics*, vol. 52, part 1, Eric Bedford, John P. D'Angelo, Robert E. Greene, Steven G. Krantz (eds.), Amer. Math. Soc., Providence, Rhode Island, 1991, pp. 245–249.

[157] S. Yamashita, "Functions of uniformly bounded characteristic on Riemann surfaces", *Trans. Amer. Math. Soc.*, vol. 288, 395–412, 1985.

SUBJECT INDEX

A

B

C

D

E

F

Juhani Riihentaus

G

H

I

J

K

L

M

N

P

Q

R

S

T

V

W

www.ingramcontent.com/pod-product-compliance
Lightning Source LLC
Chambersburg PA
CBHW041713210326
41598CB00007B/630